Naturally Late

New Critical Humanities

Series Editors:

Birgit Kaiser, Associate Professor of Comparative Literature, Utrecht University;
Timothy O'Leary, Professor of Philosophy, University of Hong Kong;
Kathrin Thiele, Associate Professor of Gender Studies and Critical Theory, Utrecht University

This series has a twofold aim: to apply the best humanities methodologies in the study of a wide range of issues that are of global, transnational significance; and, at the same time, to develop a self-reflective critique and transformation of those methodologies themselves. The books in this series will contribute to a re-imagination of critique itself; its powers, its strengths, and its possible effects.

Titles in the Series

Untranslating Machines: A Genealogy for the Ends of Global Thought, by Jacques Lezra
Naturally Late: Synchronization in Socially Constructed Times, by Will Johncock

Naturally Late

Synchronization in Socially Constructed Times

Will Johncock

ROWMAN &
LITTLEFIELD
INTERNATIONAL
London • New York

Published by Rowman & Littlefield International, Ltd.
6 Tinworth Street, London SE11 5AL
www.rowmaninternational.com

Rowman & Littlefield International, Ltd. is an affiliate of
Rowman & Littlefield
4501 Forbes Boulevard, Suite 200, Lanham, Maryland 20706, USA
With additional offices in Boulder, New York, Toronto (Canada), and London (UK)
www.rowman.com

British Library Cataloguing in Publication Information
A catalogue record for this book is available from the British Library

ISBN: HB 978-1-78661-193-2

Library of Congress Cataloging-in-Publication Data Available

ISBN: 978-1-78661-193-2 (cloth : alk. paper)
ISBN: 978-1-78661-194-9 (electronic)

Contents

Acknowledgments

I am drawn to perspectives that destabilize the idea that anyone, or anything, is ever exclusively responsible for any occurrence. Nevertheless, despite my fondness for appraisals of dispersed involvements, there are individuals whose contributions to this work I wish to distinguish.

I am appreciative of the editorial insights of Birgit Kaiser, Kathrin Thiele, and Timothy O'Leary, in directing me to what worked best about previous drafts of this book. Our interactions have always been useful and friendly, with particular gratitude to Birgit, as lead editor.

As these considerations were first forming during my doctoral studies, Vicki Kirby's relentlessly critical angle on the nature/culture question, and input regarding my perspectives, provided guidance and inspiration. I have also benefited from Melanie White's mentorship in this period on the rigors required in Bergsonian scholarship.

To my mum and dad, without whose support I might not have been able to explore philosophical questions, thank you.

Finally, sections of chapter 4 originally appeared in my article "Modifying the Modifier: Body Modification as Social Incarnation," in *Journal for the Theory of Social Behaviour* (Johncock 2012). It is with permission from Wiley that sections from that article are reproduced here.

Likewise, sections of chapter 5 originally appeared in my book chapter "Climate Change, Socially Synchronised: Are We Really Running Out of Time?" in the edited collection *What If Culture Was Nature All Along?* (Johncock 2017). It is with permission from Edinburgh University Press that sections from that book chapter are reproduced here.

Introduction

The Timing of Times

> Time is everywhere, yet it eludes us. It is so deeply implicated in our existences that it is almost invisible.
> —Barbara Adam, *Time and Social Theory* (1990, 9)

> Why shouldn't I kill time? Time will eventually kill me.
> —Michael Flaherty, *The Textures of Time* (2011, 134)

THE TIME BEFORE HUMAN TIMES

Before humans, time already was. Then there were humans. Then humans made representations of this already existing time. Then there were human, socialized or culturalized, times.

Suppositions of this sequence inform ordinary everyday, and academically theoretical, conceptions of the human relationship with time. For advocates of this interpretation, what is often installed either intentionally or inadvertently is a characterization of time as an objectively separate, preexisting source, in the universe, into which humans are born to live a finite period. This sense of time's untouchable externality informs the feeling that time is an ever-present, oppositional force. Here I refer to those moments when time appears to be "against us," from trying to finish enough work in the day, to racing to an appointment, to catching the last bus. Time looms large, yet there is often not enough of it.

These experiences indicate how, of course, we each usually "live time" not with its "universal nature" in mind, but instead immersed in the socially convened symbols to which we refer as time. Despite the fact we instinctively refer to such socialized symbols as time, the perspective found in the opening thoughts assumes a time that sits inalterably behind the socialized clocked and calendared representations that it conditions. Given the apparent inaccessibility of this primordial timing, which is already of the universe, its potentially objective function for human representations of it emerges. No human can change the tempo of the planetary movements through which the componentries of our socialized forms of time seem to develop. Collectively we can merely represent this temporality, which appears to always already be of the universe, in myriad ways. Once there is agreement within a population about a particular representation of time's rhythm, the presumed separation of the source of that rhythm from all humans dictates consensus about time's passage. If a social system has agreed upon a particular representation of time, which uses, for example, seconds, minutes, and hours, subjects using synchronized clocks based upon this representation will agree when each point in time arrives. This regularity with which subsequent times occur seems to suggest a singularity about time's natural rhythm, the source of the tempo of such representations, that is outside human interference or reference.

By "natural" here I mean, and will continue to define "natural time" as, the temporality inherent to celestial, spatial/physical patterns, including the Earth's passage around the sun, the transitional rhythms of day and night, and so on. Implicit to any such attention afforded to the relations and distinctions between natural time and what, in this inquiry, will be called "social time," is an awareness of the assumption that the natural, celestial time, of material, planetary movements, is objectively separate from the socially structured, human, representations of it. In this guise, human representations of time are believed not to affect, nor to alter, natural, celestial time, but instead to distantly describe and reflect it. As this assumption, or this condition, becomes apparent, so also does the interpretation that differing individual and social versions of time merely provide contingent, perhaps even "warped," experiences of that singular, objectively separate, natural reality of time.

The notion that we live socially according to a contingent representation of what time actually, naturally is, can be a difficult supposition from which to break. It is easy to accept that we are each born into an already temporalized universe, whereby the source or origin of time must permanently preexist any individual, or collective, human impression of it. The twelve-hour analogue clock, and our individual experiences around it, did not always exist, but the planetary and natural temporality on which the clock and its associated practices are based seems to have had a perennial place in the world.

It is in considering the impression that there are individually experiential, and socially structural, versions of a naturally preexisting time, that we can discuss the apparently *constructed* status of the socialized time structures that we live. In highlighting this distinction between the timing of times, questions that emerge concern to whom such a study might be relevant, and why it should matter to them. What is it about the conceptual distinction between natural time and social times that warrants investigation?

I will respond to such queries by firstly noting that in a rudimentary regard, the distinction of natural or actual/real time, from social or representational versions of time, should contradict every reader's personal investments in experiences of socially represented time. I identify this possible contradiction on the basis that such experiences with socialized time structures have a genuine reality for an individual, and even for a population, beyond the status of a purely contingent *mis*representation of what time actually is. While exploring this tension is one reason I feel compelled to conduct this study, there are further motivations and justifications to be identified. The remainder of this introductory chapter begins to clarify to which kinds of readers, and why, an interrogation of the conceptual division between real and representational times matters.

In considering such a division, my approach is not to primarily focus on, nor to document, the various human experiences with time. As will be shown, there are already numerous, empirically oriented, works, particularly in the social sciences, that are exclusively occupied with such a method. Instead, I am interested in the way that what I call "philosophies of time" reflect upon the conditions that are embedded within methods of time-analysis, particularly in their own methods, which might perpetuate, or alternatively destabilize, the conceptual division between naturally real and socially constructed temporalities.

It is due to this philosophical content that an anticipated audience of this book unsurprisingly comprises philosophy students and scholars. As a result, however, of this philosophical reflection also being directed toward empirically developed methods and conclusions regarding time-analysis found in the social sciences and cultural studies, it will soon be apparent why this book is also relevant for readers coming from a diversity of such fields. Furthermore, given how everyday themes and experiences of time navigate this work, the intention is also that readers who are coming from outside academic circles will feel accommodated in this evaluation of time-theory.

The reason for analyzing how fields outside philosophy handle the question of time is attributable to the interest they exhibit in the aforementioned "constructed status" of different social time structures. The social sciences in this regard typically explore how individuals and societies variously characterize, symbolize, normalize, and structure time. Forms of anthropology indeed assume a certain responsibility in guiding inquiries that seek to build an awareness of social time's differentiations around the world. Anthropologies take this role by asking how the diverse characterizations, symbolizations, and normalizations, of time, inform, and are informed by, cultural difference. I assert that it is through developing a dialogue with sociologies and anthropologies that are concerned with how social times manifest, that an appreciation can be engendered of the interpretation that there is a variability of social time, in contradistinction to a singular, natural time.

The sociological and anthropological identification of the link between cultural difference and time structure is, it could be said, worthy of an investigation in itself. In fact many of the socially scientific voices that are encountered in the initial stages of this book come from projects of that specific intent. It is, however, necessary to indicate from the outset that in this undertaking, such discussions are used as the basis of an inquiry that I anticipate to hold a different kind of rigor. This different rigor comes from a focus on how philosophy reflects on the status of social time generally. My project is thus not simply an appraisal of what is different between cultures' or populations' socialized structures of time. Nor indeed could this be said to be a work that falls somewhere between the imperatives of social science, and of philosophy. Instead, by studying how critically aware certain philosophies are of the constructed status of social time, an inquiry is developed that not only engenders self-reflective pauses for the philosophies of time concerned but also offers interdisciplinary insights to the relevant social sciences. Rather than being overly concerned with acknowledging the empirically reported particularities of cultural differences in social time structures, I am interested in how all such social time structures are positioned in relation to a supposedly, separately natural, governing temporality. Moreover, this will ask, how is it that by revealing the conditions underpinning the relationship between natural time and social times, the distinction of the latter from the former becomes more precarious than precise?

Given this theme of attending to suppositions of a sequential distinction between natural reality and social representation, it might come as no surprise to philosophically experienced readers to learn that two of my primary engagements are post-structural and phenomenological. These domains provide methodological frameworks by which the conceptual opposition between a prefigured natural phenomenon, and subsequently human phenomena, can be interrogated. Furthermore, both of these perspectives facilitate questions about whether the components of before-and-after sequences actually share exclusively forward-moving, linear relations. As we will explore, these are two reasons I refer to the methods selected from these fields, along with Henri Bergson's perspectives, as "philosophies of time."

In speaking to readers coming from an academically philosophical background, the way these philosophies are approached in this book is designed to appeal to both students and to scholars. This work does not simply offer an opportunity for students to become better appraised with these philosophies, even if such familiarization might be a part of its overall value. More importantly, readers undertaking formal studies in these areas of philosophy will find an interest in the application of such philosophical methods to contexts beyond the strict boundaries of the theories themselves. While the style of these theoretical engagements is designed to be accessible to student-readers with a relatively restricted understanding of the associated philosophers/domains, this does not, however, preclude an audience of more philosophically experienced readers. By looking at time analysis through philosophical reflections on method, this book goes beyond providing a set of exegeses of well-known literature to a market already replete with such work. Rather, in applying each philosophical position to a case study with which it is not typically associated, and in which distinctions between natural and social temporalities emerge, novel dialogues with such philosophies are produced. From this discussions about the critically interventionist utility of established philosophies should become an appealing focus for both students *and* scholars of such philosophies. This will be particularly evident when considering how these philosophies reflect on the timing of their own philosophizing, in relation to the reality about which they each philosophize.

How such philosophies unpack the theoretical preconditions of their methods, and the extent to which they destabilize the presence within such methods of the constructionist polarization of natural time and social times, will guide us in revealing our own presumptions regarding temporal separatisms. Readers from all backgrounds can in this sense feel invested in this work. Our relation to the developing dialogue is not only as witnesses to an illumination of philosophical methods of time analysis, but is also as participants who are impelled to consider our ingrained perspectives about time.

By being receptive to how the distinction between natural and social times emerges not only in philosophy, but in sociology and anthropology as well, the aforementioned interdisciplinary audience for this book includes social science students and scholars. Consistent with the earlier qualification that this is a philosophically oriented work, rather than that which falls between philosophy and social science, it is worth noting that the integration of material from the social sciences is not a driving feature of the impending argument. Instead, socially scientific evaluations of time are used as a domain to which the philosophical intervention responds. It is anticipated that a social science readership coming from outside philosophy will be interested in the patterns that are identified in certain socially scientific evaluations of time. More importantly though, I appeal to such readers' critical inclinations, and offer a specific investment in this undertaking, via the destabilization of socially scientific, constructionist theses, which divorce singular, natural time, from what are said to be relatively produced, social/cultural representations of time.

This "constructionist" element refers to the interpretation that there is an insurmountable gap between; (a) the representation or idea of reality that is developed through various social institutions or collective populations, and (b) the reality itself that is being represented. This impression, as we will explore, is not restricted to one particular type of field, theory, or practice. When I discuss "social constructionisms," therefore, what is being apprehended is not a separate field of enquiry with its own delineated borders. Rather than intending to separate out social constructionist thought from established literature, what I am interested in recognizing are aspects of the fields already discussed that embody or exhibit characteristics of the division it supposes between a natural reality, and socially constructed representations, of time.

Social constructionism is interesting methodologically, in that attending to signs of it can evidence a discipline's self-awareness, and consequent self-examination, regarding whether its own productions form part of a fabricated experience that is separated from the reality it is trying to describe. The theme of social constructionism is in the first chapter to be framed by a well-known characterization of it from Ian Hacking. In *The Social Construction of What?* (1999), Hacking advises that a central mandate of social constructionism is to highlight that the social structures that guide any particular human experience were not "inevitable." While other positions regarding social constructionism are considered in conjunction with this, Hacking's definition is important to raise now. Through his focus on the theme of inevitability (or lack thereof), we encounter the social constructionist argument regarding the difference between natural and social phenomena. This requires that while the natural realm is governed by necessary, physical laws, conversely social structures, and theories or ideas about such social structures, could have manifested in any number of relatively accidental ways. Regarding the topic at hand—time—a social constructionist methodology could duly surmise that it was not inevitable that time was conceived in any particular symbolic form (such as the twelve-hour analogue clock). Rather, such representation is contingent upon numerous contextual or political factors and could have taken myriad other forms. The very *idea* of time is socially constructed in this view, as are the forms via which a time-consciousness becomes collectively installed.

Why an examination of the construction of this idea matters, or warrants an investigation, concerns not only its relevance to scholarly impressions of time, but also to how we each live with time. When the time on a clock, that is perhaps found on your watch or smartphone, reads a slightly different time to that of your friend's clock/watch/smartphone, the ensuing conversation about whose clock is "right or wrong" reveals what is at stake in the social constructionist dilemma; that representations of time are variable and potentially fallible. Our lived experience with these representations of time is generally something that we take for granted, in that we often live as though the clock *is* the only time. However, it is during a moment of temporal disagreement such as that just mentioned, where representations are re-presented as just that—representations—that we might appreciate the clock to be but one of many constructions of another kind of time. It could also be that in these moments we distinguish certain clocks as being closer to, or further from, an underlying actuality of time.

Beyond individual experiences of asynchrony, it is when different, socially developed time-structures are considered to be more, or less, distanced from time's naturally real state, that I believe a politics emerges regarding how developed, or undeveloped, the cultures associated with such time-structures are deemed to be. I expect that the critical attention that is afforded to this point in the coming debate will appeal to the sensibilities of readers to whom the conceptual subordination of certain cultures around the world matters. That a philosophical response to this ramification is one focus of this work evidences my motivation to open the interventionist capacities of the methods that are to be engaged. Such critique duly targets contexts where the constructionist distinction between natural time, and social times, has real and durable effects.

WHY LATENESS? WHICH LATENESS?

The conceptual separation of natural time from the plurality and relativity of socialized times is, I believe, able to be explored through a particular trans-situational theme—lateness. The culturally relative constitution of social time can be identified via lateness protocols, which differ remarkably between societies. As becomes apparent in the sociological and anthropological, empirically produced, claims that are presented in the initial phase of this inquiry, numerous cultures might use the same symbolic structures to frame temporal agreement and synchrony between their respective citizens. However, such cultures or societies can vary dramatically in how they regulate individuals regarding collective time-agreements. Or in more simple terms, different cultures have different rules, definitions, and punishments, in relation to being late. In accordingly responding to a possible query of why "lateness" is the theme being used for my examination of the relation between natural time and social time, I assert that lateness is a parameter by which we can appreciate, and interrogate, suppositions of the contingent relativity of social time's protocols and processes.

There is also a practical criticality to the use of a theme such as lateness, in the way that it accommodates all anticipated readers. "Being late," or trying not to be, is an aspect of modern life that would be familiar to many, if not to all, readers. Furthermore, beyond the context of social lateness, it is through case studies of the timing of bodily changes, and of responses to climate change, that the philosophies to be featured reflect on what *lateness as secondariness* means for their impressions of natural and socialized temporalities. From this philosophical introspection and application, all readers will be impelled to consider which assumptions they might carry regarding time. Philosophy, in unpacking the time-perspectives of three of its methods, in this sense will undertake a similar kind of self-reflection as is encouraged of the reader. While this work is philosophically navigated accordingly, a prior expertise in such theory is not necessary however. Instead, the contexts of what might be secondarily late about various cultural productions and behaviors in social, bodily, and climatic settings offer a broader relevance of such engagements than to a restricted, academically experienced, audience.

When reviewing certain cultured conceptions of lateness in the early stages of this book, it is considered how lateness is portrayed as an important component of social and commercial relations in typical, capitalist, social time structures. Complementing this analysis are accounts of social structures in which lateness supposedly does not figure as prominently. While many of these latter examples are taken from non-Western cultures, the intention is not to polarize Western from non-Western time scales and protocols, as though either could be homogenized. My interest in the differing regulation of social and commercial lateness around the globe indeed goes beyond a comparative exercise that anthropologically maps where various cultures sit in relation to others. While an acknowledgment of the particular qualities of the cultural/social times that are included in this study is pivotal, such qualities are raised to serve a focus on how *any* socially structured time, and therefore social time *in general*, is conceived as being social because of its perceived distance from a natural temporality that it might represent, appropriate, or abstract. The differences reported by the empirically oriented social sciences between time structures or cultures are in this endeavor not necessarily as important as is the philosophically derived concern with the general conditions that emerge when social time's relativity is interpreted to distinguish it from natural time. It is through the theme of lateness that we can begin to navigate theoretical positions on the contextually constructed conditions of social time's variations, and determine what these variations highlight regarding our, and philosophy's, conception of social time's status in relation to natural grounds. As a result of this approach, readers from a philosophical background might find interesting the considerations of the utility of philosophical methods of time-analysis.

As earlier indicated, this variability of social time needs to be considered at both a structural level and at an individual level. While there are certain, objective features about the structure by which social time operates for a population, the particularities of one's lived experience within such a structure become overtly evident in terms of lateness. The jurisdiction of a social structure of time seems to be all-encompassing for any individual living with it, but nevertheless individuals can be asynchronous with such a structure when their tardiness contradicts an otherwise expected timeliness. From this we can ask whether such lateness represents a *breakdown* of what is collectively demanded about social time, or even a transgression of that time by the subjectivity of individual time. At play in such considerations of "temporal breakdown" are notions of sequence and simultaneity, and whether socially structured protocols simply *precede* individual temporalities. Lateness regulations here provide an insight into what is perhaps the typical impression; that there is a socially structured temporality before there are subsequent, individualized, versions of it, that each of us live to varying degrees of coherence and synchronization with our fellow citizens.

This query speaks to the tendency, both at anecdotal and theoretical levels, to want to discuss lateness in terms of what is subjective about time. In considering supposedly contingent constructions of time, what becomes relevant is how in the same time-context, qualitative variances are shown regarding what social time *feels like* to different individuals. A most important question in this sense regards whether an individual's lateness symbolizes a defiance of social time, whereby an individual could be interpreted to be able to intentionally move to a rhythm that is separate from that which has been collectively coerced. This project is not empirically exploring subjective time-experiences, nor is it overly invested in the abundance of such experiences. Nevertheless, in philosophically interrogating the social constructionist division of natural and social times, empirically derived positions are targeted that assert that a diversity of subjective time-experiences indicates a distinction between time as it is contingently individually lived/represented, versus a real time on which all such representations are based.

Implicit to social constructionist arguments positing the contingency of structures of social time is the recognition that established structures favor (and are designed to favor) certain populations and demographics while marginalizing others. Despite, therefore, the earlier clarification that the focus is on the general conditions of social time (rather than on the relative position of any particular culture's social time), it is incumbent upon the impending study to consider whether the philosophical reflections that manifest embody a certain emancipatory potential for marginalized demographics. This concerns the capacity for a philosophical method to intervene where institutionalized, invisible, oppressive social time-structures are dogmatically instilled to the detriment of certain population groups. The cultural relativization of social time(s) that constructionist conceptions offer seemingly has egalitarian intentions. It is designed to illuminate why no social time is the global, default time, where "global" refers to the differently cultured ways in which humans inhabit the Earth. However, what I want to ask, via philosophical reflections on the conditions of time-analysis, is which complications are presented by the constructionist relativization of social times if the status of any such time relies on its differentiation from a natural, celestial, time?

The philosophies in which I identify reflections on the comparative parameters of the above argument are presented in three primary engagements. The dialogues that develop with Derridian post-structuralism, Merleau-Pontian phenomenology, and Bergson's theory of time, explore how philosophies reflect upon the conditions underpinning their approaches when confronted with apparent sequences from "naturally material or spatial" to "culturally abstract" temporalities. Their relevance to my inquiry can, in this sense, be attributed to the suspicion that is identifiable in the work of each regarding suppositions of straightforward before-and-after temporal relations. This suspicion is evident not only in terms of the separation of natural phenomena from social phenomena, but also regarding suppositions of the disjunction between reality and representation. In each method I find a curiosity regarding how the representation of a preceding real could possibly manifest that either lacks the material nature of, or has cultured properties in excess to, that real that is represented. Their inquiries in turn encourage me to think regarding the question of time, that if both the real (natural time), and the representation of the real (social time), materially occur in the same universal realm, then what clear distinction can we draw between the origin of the constructed phenomenon, and the origin of the phenomenon on which such a construction is based?

Jacques Derrida's post-structuralism provides a valuable perspective on sequences between natural and social origins that he identifies in aspects of anthropology and philosophy. Specifically, Derrida asks us to interrogate the presumption that there is a pure, externally prior, state of nature that is antecedent to representation, or that was not already itself a kind of representing. In a similar regard, the value of Maurice Merleau-Ponty's phenomenology for this project concerns its reconfiguration of the Husserlian eidetic reduction. The eidetic, or phenomenological, reduction attempts to suspend (bracket) judgments in order to identify what can be removed from object experiences without eliminating what is "essential" to them. Merleau-Ponty's inquiries into the notion of what is essentially or objectively separate about phenomenal experience, outside and prior to our perceptions of such phenomena, are identifiable as being of assistance to my interest in exploring the presumption of a preceding temporal purity from which subsequently contingent temporalities are distanced. It is with such considerations that we likewise find the worth of Henri Bergson's reflections about whether the "real of time" is ever without the kind of representation that is attributed to socially temporalized perspectives. By considering the interplay between Bergson and Merleau-Ponty on these matters, the temporality of theory and critique can also be debated, potentially muddling with the supposed sequence between a prior, phenomenal world, and consequent, socialized reflections on that world.

As the question of the temporality of theory is raised, so questions must be asked regarding at which point a theoretical body of work is distinguishable from the material realities that it engages. Exploring the temporal relations between theory and reality has ramifications for our understandings of, and relations with, method. If we, as physically material, socially structured, creatures are to some extent implicated in the theory and philosophy that we produce and critique, and vice versa they in us, are materiality and theory distinctly separate? By interchangeably discussing bodies of humans and of theory, what manifests are possible porosities requiring attention between the materialities and the temporalities of corporeality, social structure, and philosophy.

There are various reasons that this inquiry into philosophical perspectives on the conceptual separation of natural time and social time, through the prism of lateness, ventures into the context of the human body. Consistent with the just identified concerns about assumptions of a prior, or an external, purity, an analysis of the politics of bodily change illuminates aspects of the sequence in which a natural, unadulterated temporality is presumed to precede socially fabricated temporalities. Here I draw a correlation with the context of socially constructed forms of time, in which differing cultural standards around the world about lateness and synchronization complement the view that there are subsequent human representations of a natural timing. Likewise with the human body that is materially reshaped, reconditioned, and reconstructed, via socially informed processes (technological, medical, beautifying, etc.), a distinction emerges between natural, and socialized, corporeal temporalities. In this sense the timing of the modified body is used in this project as a case study, parallel to the case study of social time protocols. Central to considerations regarding the socially constructed body is the acknowledgment that if this project is interested in social or socialized time generally, and this is a time that refers in some way to *human*, social time, then the embodied nature of the human demands an attention on the intersection between time and the body. Indeed a complete absence of discussion regarding the *human* body, given this project's attention to specifically *human* social phenomena, would be glaring.

Socially constructed bodies, manifesting from practices including exercising, tattooing, dieting, body piercing, tanning, and plastic surgery, seemingly produce distinct "before bodies" and "after bodies." Such practices, emerging from social contexts, contribute to the production of changing, materially spatial, bodies and bodily limits.[1] Bodily limits represent not just the visible points where a physical, corporeal entity begins and ends. Rather, such limits, such temporalizations, also express social contexts. These "social bodies" become distinguishable according to the norms, meanings, and interactions that mark the finality of one social form and the commencement of another. Despite these inherent limits, there is nevertheless an implication between social contexts, between social bodies, that confounds any simple sense of separateness. Yet even taking into account this continuity, when it comes to the relation between natural temporality, and its representational social forms that are said to follow, a break or abyss is posited by the constructionist perspective in which the representational unsuccessfully tries to access the essential truth of the naturally real.

What I wish to ask as a result is whether, via deconstructive reflections on theses about human corporeality, socially constructed bodies are considered to arrive late[2] to an already occurring, natural corporeality. "Late" might seem like a curious term to use in such a context. However, when notions of origination and purity regarding the body are given greater attention, impressions of the "untimeliness" of social intrusions upon natural phenomena become evident. A study of perspectives about the social modification of the body is duly positioned as an inquiry into the perceived secondariness of social representations of an already occurring natural reality. Moreover, it is via an examination of bodily temporalities that we see the suitability to a project such as this of certain thinkers within the history of phenomenology. In particular, Merleau-Ponty's body-focused response to his phenomenological predecessors, Edmund Husserl and Martin Heidegger, attends to the potential determination of the corporeal limit in relation to the world in which it manifests.

In opening the discussion about bodily limits and lateness to this context of theoretical or philosophical work about the body, so I wish to raise the issue of theory's or philosophy's materiality. What I am intrigued by here is the relation to the materiality of the body, of theories or philosophies about the body. Is a straightforward, natural-social sequence exhibited, in which firstly the real body naturally exists, before socially and culturally informed theoretical representations of the body arrive late on its already established scene? This appreciation of secondariness, which I believe again distinguishes natural time from social structures of time, and the natural body from the socially modified body, conceivably also commands a distinction between a material reality, and the theory of that reality, that I wish to address. The timing of theory is in this book a third case study or context, complementing discourses about body modification practices, and assertions about the globally plural forms of social time, through which the timing of social phenomena in comparison to its natural antecedent is reviewed.

We often in fact refer to theory as being bodied, in terms of "a body of work," even if we do not typically extend the sense of that body of theory to being literally, physically, materially embodied. But what if we were to raise that literal, physical possibility in this undertaking? What would the ramifications be for suggesting that bodies of representation, social and philosophical in constitution, might not be straightforwardly separate from, subsequent to, or lack the material properties of, the bodily realities on which they are based and to which they direct their attention? Why such a question about the materiality of theory would matter circles us back to the original consideration of natural time and social timings. The compulsion we feel to characterize human activity as culturally constructed appropriations of the naturally real indicates assumptions of a secondariness pervading everything about our social existences. My day-to-day consciousness, my lived experience, normalizes such compulsions like anyone else's. Yet when I reflect on this logic as I do now, of an apparent gap in the reality-representation temporal sequence, I am confused. From where, after all, does representation originate, if not as a mode of the reality from which it is supposedly separate?

In exploring these types of problematics, it is appropriate to signal that I am continuing a line of inquiry inaugurated by a series of thinkers who question the assumption of a nature|culture divide. My specific uneasiness with the conceptual separation between natural time and social time can in this regard be located within the broader intellectual impetus of the likes of Derrida. By deconstructing the separation of nature from culture that he identifies in the anthropology of Claude Lévi-Strauss, for example, his work frames the kind of trajectory that my inquiry takes (explaining why Derrida is engaged at length in chapter 4).

As questions of the division of nature and culture develop with an eye on particular political ramifications, so my interest in how the separation of natural time and social times affects different populations can likewise be situated within a legacy established by inquiries such as Donna Haraway's "Manifesto for Cyborgs" (1985). Haraway's recognition of postmodern, cyborgian, organic-machinic identities marks an intervention against patriarchally institutionalized divisions that are held to be responsible for ascribing to women a biologically determined, acultural constitution. This kind of inquiry is in fact inaugurated well before Haraway's work, for example in Sherry Ortner's critically inclined "wish to see genuine change" (Ortner 1972, 5) in the structure that assumes that "female is to male as nature is to culture" (1972, 5).

While these interrogations of the nature|culture split inform a consciousness that makes undertakings like mine possible, I can more precisely situate my project within a recent genre of thinkers who reconfigure the nature|culture split according to assertions about a conceptual division between reality, and socially constructed representations of reality. Vicki Kirby's contributions are notable in this regard, whereby in *Telling Flesh* (1997) she contests the "quite definite separation of representation and information from physical life" (Kirby 1997, 134). This perspective influences my specific interest in what I view to be the similarly perceived difference between the temporality of nature, and the human representations of that temporality. The "physical" aspect to life that Kirby mentions (and that she observes is problematically believed to be separate from the humanly representational life), correlates with a "natural time" that I interpret to be regularly aligned with a separately physical, celestially regular, rhythm of the universe. Natural time in such a conception is, in my view, positioned in contradistinction to its social reproductions, as a reference to the physicality and materiality of the world's "pre-representational" rhythm.

My investigation of assumptions that culture constructs, and distantly represents, the "naturally real," can also be placed within a line of thought that Susan Oyama has helped to develop. For Oyama, we must oppose the notion that the dogmatism of biological determinism can be managed with the "antidote" of social constructionism, given that doing so simply regurgitates the nature-nurture binary (Oyama 2000, 16–17). This is a point that Kirby also makes (Kirby 2006) in her critique of aspects of Judith Butler's impression of the performativity of gender. I do not wish to exclude either nature or culture from time's reality. In therefore examining how philosophy-as-culture critically reflects on its methods and preconditions regarding the nature|culture question about time, I am conscious of working, as Haraway has suggested, from "inside the belly of the monster" for "[w]ithout the nature/culture split, how can nature be reinvented?" (Haraway quoted in Penley and Ross 1991, 7).

Where Haraway elsewhere says in an interview that she is "neither a naturalist, nor a social constructionist . . . it is not nature. It is not culture" (Haraway 2004, 330), the specificity of my focus from hers then diverges. This is because I wish to explore how philosophy reviews itself in a way time might be conceived *to be both naturally and socially constructed*, rather than neither, or only truly one, of the poles. This endeavor becomes most apparent in chapter 5. A brief part of that dialogue concerns the work of Karen Barad, who holds a certain status alongside the thinkers we have just encountered that interrogate the nature|culture divide in ways with which I share a relative allegiance. Barad's project, informed by her background in physics, asks not how closely a scientific apparatus, as a cultural artefact, can accurately measure natural phenomena. Rather, the focus is to identify entanglements between the apparatus and the nature that it explores (Barad 2007, 168–79). These considerations are similarly evoked in the moments of my analysis that are concerned with the representational apparatuses of time, such as clocks and calendars, which are often used in adjudications of social lateness. Complementarily, I attend to how philosophies evaluate methods in which assumptions of the secondariness of social phenomena, and the subjective experiences implicated with social phenomena, are connected to technologies interpreted to distance us from nature. As has been clarified, the motivation of this project is not to unpack the richness of these experiences that can be empirically reported and compared. Instead, one intention is to use existing empirically developed conclusions about the diversity of collective and subjective time-consciousnesses to illustrate that within such conclusions, real time and representational time are conceptually divided.

Lateness has been introduced here, via three contexts, as the theme through which the conditions of social constructions of time can be examined. The first context, the acknowledgment of differently cultured structures of social lateness protocols, and how they are signified in comparison to natural time, is of primary importance. Secondly, through the context of the socially modified body, it can be asked whether such socially constructed corporeal transitions or temporalities arrive *late* to an already originated and established natural phenomenon. From this the question is raised about the timing of natural reality generally, in relation to that of the culturally produced representation or version of that reality. The timing of theory has, in this latter sense, been flagged accordingly as the third context of this book.

These three contexts, in facilitating discussion regarding philosophical reflections on the temporal division between reality and representation, are to be complemented by a fourth context or case study in which the focus on the theme of lateness takes on an overtly existential edge. Straightforward impressions of the constructionist separation between natural and social phenomena are at stake when considering the advent of, and discourses concerned with, climate change. As with debates about the late intrusion to the natural body by socially informed modifications, relations between the world's ecology, and industrialized human societies, can be investigated through the presumption of a secondary, unnatural, social incursion to a natural scene. The additional and perhaps defining relevance of this context to the theme of lateness is how it highlights and responds to the demand that humans are too late, that they are "running out of time," to save the planet from ecological and environmental catastrophe. In an era in which a public consciousness concerning environmental responsibilities and the human relationship with climate factors proliferates, the use of such a context is designed to engender a further level of investment in this book's philosophical considerations from all readers.

The temporality between human, social realms, and a material, natural realm, must duly be considered in light of the accusation that humanity's collective response to climate change might be happening too late to address what is most perilous about current states of ecological sustainability and environmental survival. The existential ramification in this discussion is that we could not only be too late to prevent irreversible damage occurring to the planet, but also that the species could end up being entirely "late," in the mortal sense, if government action and consumer behavior does not shift in time. As we consider how phenomenological reflections on temporality are positioned to engage climate change discussions, the politics between a natural temporality and industrialized, socialized temporalities, become pivotal.

These inquiries require a reconsideration of the composition of cultural products, such as philosophies, texts, and theories, with which we representationally engage a materially changing world. What is crucial is our impression of the timing of human, cultural, and/or social, production in relation to the world toward which such production is directed. Implicit to these considerations is the question of whether philosophies of time reflect on an already existing, material phenomenon, or somehow internally participate in the material production of that phenomenon. It is during the points in this book concerned with this question that philosophy's self-reflection on the nature of its own productions becomes most explicit.

THE SOCIAL, MATERIAL, SITES OF INTERROGATION

The immediately preceding summary raises the need to distinguish between the primary investigation of this book—concerned with the social constructionist distinction between natural time and social time—and the supporting contexts that serve that primary investigation. The primary investigation has two steps:

1. the identification, through an engagement with sociologies, anthropologies, and even philosophies, of time, of the condition that is implicit to social constructionist theories of time, in which natural time and social time are conceptually separated. This condition is to be referred to as the "social constructionist time condition."
2. the response to this condition, which considers whether philosophical reflections on philosophical methods concerned with questions of time, can reconcile (rather than separate) natural time and social time.

To serve this primary investigation, the supporting contexts, each exemplifying a politics of secondariness between anterior natural time and posterior social times, are engaged:

a. the conception of the global relativity of socialized times and regulations regarding lateness, in comparison to the singularity of a natural time which social times represent;
b. the interpretation of the subsequence and divergence of socially modified bodies, in relation to the natural body from which they derive;
c. the sense that theory, as socially/culturally produced representation, occurs externally to the established natural reality about which is theorized; and

d. the impression that industrialization has distanced humans from nature, and the fear that it might be too late to "save" the planet from the consequent environmental disaster.

The term "late" carries with it many negative connotations. To be described as "late" is seemingly often adverse. In a social context being late means that one has not fulfilled their commitment or responsibilities to a social contract regarding timeliness, and in an existential regard it refers to one being dead! With this latter signification of a time that governs our existence, the mortality with which lateness is invested hints at a potential futility regarding attempts to control time via the socially structured representation of it. While we live according to a symbolic, moving picture of time's passage, time itself can seem to be an all-encompassing generative and destructive force over which we have no power or jurisdiction.

This thought coheres with the earlier observation of the conventional understanding that time is not really the form in which it is represented and constructed by humans, but rather is a worldly phenomenon that humans simply symbolize, measure, and manage. It is my position that these assumptions have genuine efficacy, exacting a material purchase that cannot be ignored in terms of our lived, embodied experiences, and thus need to be accommodated within, rather than excluded from, any theoretical undertakings on time. By engaging the earlier discussed philosophical methodologies to examine, and potentially reconfigure, the social constructionist time condition, what requires interrogation is the supposition of an asynchrony between natural and social temporalities. Whether in terms of social responsibilities, embodied or intellectual cultural productions, or human existence, being late matters.

Chapter One

Social Times

Contingent Constructions?

> We accept clock time, even though such time is a social construct. As an objective fact of daily life it provides a commonly held standard, outside of any one person's influence, to which we turn again and again to organize our lives.
> —David Harvey, *Between Space and Time* (1990, 418)

CRITICAL CONSTRUCTIONS

Having outlined the key questions of this project, its particular direction should now be in focus. What I hope to offer is an interrogation of what conditions the conception of *all* social times, of social time generally, rather than attend to the cultural differences between any particular social time structures taken from around the human world. This feature of the impending philosophical excursion will accordingly provoke a possible curiosity from readers from the social sciences, for whom theses of the cultural relativization of time might be an ultimate outcome. In focusing on the theme of lateness, my intent though is not to fascinate unnecessarily over different cultural protocols regarding lateness, whether such protocols manifest explicitly via rules concerning social synchronization, or implicitly within collective conceptions about the temporalities involved in bodily purity or responses to climate change. Instead, what is being asked is what can a study, through the lens of lateness, tell us about the conceptual positioning of times that are characterized as social, in relation to a ground or reference that is deemed to be natural? Furthermore, what awareness of the differentiating conditions that are installed between social time and natural time do philosophies interested in questions about time exhibit, not only in reviewing other fields of inquiry, but also in their own methods? As we follow three philosophical domains' reflections on the conditions regarding temporality that underpin their methods, so we should be prompted to consider which assumptions we hold about time.

It is as a result of this attention on the conditions by which social time is conceptually differentiated from natural time that I come to describe this project as having certain "critical" characteristics. In considering what motivates this characterization of criticality, a discussion can be developed regarding the interventionist qualities that are inherent to this work. By this I refer to the intention, already outlined in the introduction, to not only identify but also to destabilize, or reconfigure, a specific condition within conceptions of social time that separates representation from reality. If this intention is locatable within the philosophies of time that come to be central to this undertaking, then it can be considered whether these philosophies exhibit a similar criticality. While this direction can and will be applied to perpetuations of a conceptual divide between natural and social times found beyond philosophy exclusively, this critical element becomes most evident in this book's attention to how a philosophy self-reflects on the conditions of time-analysis that have informed its method(s). In both applications, the question of method pervades in philosophical interrogations of empiricist suppositions of a "natural world out there." A certain appeal manifests in this work accordingly for scholars of philosophy in appraising the critical capacities of aspects of philosophies not typically characterized as having such utility.

In order to structurally frame the critical significance of this study into what I have identified as the "social constructionist time condition," an awareness of critical theory's definition of itself is useful. Attending to its self-evaluation will also set a platform for the kind of disciplinary self-awareness required of the philosophical engagements that drive our inquiries in later chapters. This project can be situated in this regard by appreciating critical theory's disciplinarily restricted, as well as thematically broad, definitions. Regarding its more restricted sense, critical theory can refer to the methods and mandates of a series of philosophers and social theorists that began to emerge in the 1930s from the *Institute for Social Research* in Frankfurt. What would come to be known as the "Frankfurt School" of critical theory—denoting a grouping beyond the actual institution—developed as a collection of scholars whose general position held that traditional theory merely intended to decipher and explain social institutions and norms. Conversely, critical theory was motivated to dismantle and rebuild such social configurations. From this, a particular theme would distinguish "critical theory" from traditional theory, that being critique's intention to liberate socially oppressed groups. A familiar reference point for critical theory's self-definition in this style is Max Horkheimer's affirmation that

> [i]t is not just a research hypothesis which shows its value in the ongoing business of men; it is an essential element in the historical effort to create a world which satisfies the needs and powers of men. (Horkheimer 1972 [1968], 245–46)

This "satisfaction" of humanity's needs and powers reflects that the interventions the Frankfurt School embodies *are the reasons* for the production of the associated theories. Knowledge production and acquisition is not a critical objective in itself. Rather, actual transformation of the situations in which humans are institutionally oppressed or marginalized is anticipated in critique's appraisal of, and consequent participation in, social life, whereby "theory never aims simply at an increase of knowledge as such. Its goal is man's emancipation" (Horkheimer 1972, 246).

If critical theory is designed to interrogate aspects of human, social structures, which are potentially oppressive, then it also represents a method that examines and dismantles what is institutionalized, or taken for granted, about social experience. It is in this feature of critical theory's definition—the interrogation and reconfiguration of social institutions whose entrenched character could be detrimental—that we encounter the first indication of how a study into the social constructionist time condition exhibits "critical" inclinations. Social time is one such normalized institution, a seemingly inescapable truth of life in a social context in which one gravitates. Matching our intentions here, David Hoy makes apparent the applicability of characterizations of criticality to interrogations into what is taken for granted about time in *A Time of Our Lives: A Critical History of Temporality* (2009). By analyzing established social structures and identifying how they covertly benefit certain populations and demographics over others, Hoy notes that via critical inquiry "we begin to suspect that social arrangements that we take to be just are in fact hiding oppressive relations of force" (Hoy 2009, 228). For Hoy, critical theory, when attending to how various races, genders, bodies, and personalities experience social time differently, is the approach that has made the greatest incursions into the "oppressive social conditioning" of time-structures (2009, 228). As our inquiry develops, relations of synchronization and lateness will emerge as forms of this "social conditioning."

Hoy's portrayal is consistent with the earlier insight presented by Horkheimer regarding the differentiation between critical theory and traditional theory. While there are prominent thinkers other than Horkheimer who are associated with this generation of the Frankfurt School, I choose to focus on Horkheimer to guide us through this initial engagement of critical theory. This is for two reasons. Firstly, it is Horkheimer who formally defines critical theory in a manner that is said to navigate the Institute's early conception of itself.[1] Secondly, the critical perspective of my project will go beyond the Frankfurt School's specificities, incorporating genres of critique offered by deconstruction and "New Criticism," to name just two. Despite this, the points of intersection between the Frankfurt School approach to critique and my inquiry into philosophies of time become most apparent in terms of Horkheimer's characterization. Part of this characterization is found in the definition offered in the 1937 essay "Traditional and Critical Theory," in which Horkheimer elaborates that rather than attempting to provide an objectively neutral discourse, critical theory embraces its own situated, and therefore self-referential, interest in its subject matter (Horkheimer 1972, 188–243). This sense of a theory's capacity to reflect on its situatedness in relation to the issue being critiqued will be of great importance when considering the timing in this book of socialized, representational, productions and constructions.

These themes are not all that are at stake in the Frankfurt School critical theory. After World War II, theorists associated with the Institute for Social Research begin to analyze the integration of the working classes into modern, capitalist societies. The escalating institutionalization or bureaucratization of what was becoming a "one dimensional society" becomes their focus, attending to how newly established industries of culture were standardizing consciousness.[2] In 1947 this motivates Horkheimer and Theodor Adorno to argue in *Dialectic of Enlightenment* (2002 [1947]) that a relative end to subjectivity is being capitalistically imposed, whereby "the individual is entirely nullified in the face of the economic powers" (Horkheimer and Adorno 2002, xvii). Twenty years later, in 1967, Adorno expands on these themes by positing that mass culture does not reflect the experiences of the people but rather is installed as a commodifying industry of powerful institutions (Adorno 2001, 98–106). This indeed leads Adorno to suggest that a key aspect of critical theory is its inquiry into the connection between goods that are considered to be useful for individuals versus what it is that constitutes culture. His response is that a labor force is not deployed simply for the industrial production of what is used for humans' actual survival, given that much of what is "classified as useful goods goes beyond the directly biological reproduction of life" (2001, 114). Industry is duly seen to have repositioned the individual as a homogenized consumer of cultured goods.

The critique of mass society's repression of subjectivity is in fact often portrayed as being most effectively articulated by Adorno's Frankfurt School peer Herbert Marcuse.[3] In *One-Dimensional Man* (2002 [1964]), Marcuse argues that by incorporating and defusing oppositional impulses within society, modern industrialization negates what would form the subjective bases for interclass conflict and social instability. In order to conversely generate industrial productivity, "class struggles are attenuated," whereby "capitalist society shows an internal union and cohesion unknown at previous stages of industrial civilization" (Marcuse 2002, 23). In Marcuse's ensuing argument that capital structures exploit labor forces while making individuals dependent on them (2002, 27), what manifests is a thesis that is no less consistent with the navigations and motivations of the Institute than that earlier encountered from Horkheimer.

Nevertheless, it is Horkheimer's originary characterization to which I return in identifying how the Frankfurt School informs features of my inquiry into philosophy's critical perspectives on time. This is because of Horkheimer's emphasis on the self-referential and interventionist qualities that he posits critical theory should embody. We will discuss critical theory's self-referential features shortly. Regarding its interventionist intentions more immediately, it is Hoy's reading that by broadcasting this purpose, critical theory does not artificially transcend the social environments that it critiques. By interrogating institutionalized stratas of dominance and oppression, critical theory can instead stand as a "resistance to co-optation by the dominant class" (Hoy 2009, 228). This of course is consistent with Horkheimer's conceptualization of the purpose of critical theory, which he describes as being more complex than a mere concern with perpetuating "the better functioning of any element in the [social] structure" (Horkheimer 1972, 207). Horkheimer's position regarding critical theory instead recognizes its liberationist aspirations by attending to the conditions by which any "element" becomes presently established. Having identified what is instilled or institutionalized, it is through the consequent interventions that disrupt and reinvent social worlds, that for Horkheimer critical theory reconfigures "the very categories of better, useful, appropriate, productive, and valuable as these are understood in the present order" (1972, 207).

While I engage Horkheimer's definition of critical theory as explained earlier, what is critical about my investigation is not entirely reducible to Horkheimer's declarations and motivations regarding the strict emancipation from oppression of certain classes or demographics. Certainly, the Frankfurt School's intention to interrogate and reconfigure, rather than to merely understand, the social present is an ethos that is embedded within the discussion through which I lead us. Furthermore, it informs a reason that all readers might care about a work such as this, which, by engaging critical perspectives on the reporting of natural and social times, allows us to question taken-for-granted social time arrangements and their inherent power structures. However, when situating where this examination of the conditions underpinning the relation between natural time and social time sits in terms of critical theory generally, the earlier mentioned "broader" sense of critique becomes relevant. In order to unpack this broad, critical relevance, it is necessary to incorporate a definition of our overall target; the "social constructionist time condition." Such a definition can be most clearly presented by firstly discussing the pivotal parameter within that target: constructionism. During this constructionist discussion, we should take note of suppositions of a divide between reality and representation, given how this informs impressions of the secondariness of socially symbolic temporalities, in relation to a "real world" temporality on which they are based.

Just as I am in the process of structurally determining the sense of criticality that permeates this project, similarly I will approach the task of defining constructionism relationally. Constructionism can be situated by integrating a complementary discussion of what could be considered its parent ideology—constructivism. The form of constructivist thought that is most relevant concerns the epistemological insights derived by Swiss psychologist Jean Piaget on child cognitive development. In *The Origins of Intelligence in Children* (1952), Piaget argues that intelligence is not indicative of innate individual traits. Rather, intelligence manifests according to a combination of biological maturation and external interactions, "entangled in a network of relations between the organism and the environment" (Piaget 1952, 19).

Given the role of "hereditary adaptations" in the organism, Piaget later describes the consequent inquiry into this conception of knowledge development in terms of a "genetic epistemology" (Piaget 1970). Piaget defines genetic epistemology as an attempt to explain knowledge on the basis of its genesis, distinguished from the "traditional philosophical view of epistemology" that only offers a "study of knowledge as it exists at the present moment" (1970, 1–2). Rather than being interested in "knowledge for its own sake" (2), Piaget explores how humans come to produce, or *construct*, knowledge and meaning. The basis of this construction is established in *The Origins of Intelligence in Children* and is said to be a set of mental schemas, as frameworks of knowledge, that individuals build experientially in mentally representing the world. Knowledges derived through new experiences are only given meaning in accordance with one's existing knowledge schematic. Given that "every schema is coordinated with all the other schemata," no knowledge of the world is developed outside one's already cognitively established, impressions of that world (Piaget 1952, 7). Piaget hence gives us an insight into how the component parts of a knowledge-process that can be defined as a "construction" co-develop.

The systemic connection between one's biologically hereditary, cognitive maturation or development, and their environmental interactions is for Piaget attributable to two processes. Firstly, he posits that individuals "assimilate" new experiences with the old experiences that they schematically hold or "know." One's existing schemas drive their management of, and education about, a new object or situation. Secondly, and of significance to our understanding of how knowledge develops for the constructivist, Piaget acknowledges the consequent plasticity of what we think we know to be true. The knowledges that one "has" are not permanent, untainted, cognitive concretions but ongoing collegial processes of "modification" (Piaget 1952, 6–7). What he then describes as the "accommodation" of new knowledges occurs when one's existing knowledge schemas are not adequate for the world that is experienced and thus need to receptively adapt. Without venturing into aspects of Piaget's thesis that transcend our needs, the important point for us is that the new information the individual obtains is integrated into, and *constructed with*, one's existing knowledge structures. In taking this point, we are beginning to open the question regarding the temporality between reality, and the constructed knowledge of that reality.

Such an outlook would eventually inform pedagogical debates regarding learning and teaching practices. However, just as my interest in social time is not overly concerned with the measurable, sociological or anthropological differences between social time structures, but rather is focused on the conditions by which we come to know time's social status, correlatively Piaget is not intent on applying his theory to reveal the differing measures of how well children count or solve problems. Instead, as with my undertaking, he wants to explore the conditions of representational knowledge, the "'principal categories' which intelligence uses to adapt to the external world—space and time, causality and substance, classification and number" (Piaget 1952, 8).

It is when these basic tenets of cognitive constructivism are applied to pedagogical theory that insights emerge for our currently unfolding sense of constructionist impressions of knowledge development. One such prominent application comes via Seymour Papert, a student of Piaget's. Papert continues Piaget's interest in the construction of knowledge structures. However, rather than exploring issues of cognitive development, Papert focuses on learning processes as an interactive production in which the individual participates, instead of simply observes. As he states in a famous proposal to the National Science Foundation,

> The word constructionism is a mnemonic for two aspects of the theory of science education underlying this project. From constructivist theories of psychology we take a view of learning as a reconstruction rather than as a transmission of knowledge.[4] Then we extend the idea of manipulative materials to the idea that learning is most effective when part of an activity the learner experiences as constructing a meaningful product. (Papert 1987, 2)

The participatory component in knowledge construction is emphasized in order to convey how the individual learner constructs their own sense of meaning. Meaning is not merely transmitted as a timelessly true fact but is developed through practical involvement. As Papert clarifies though in the "Situating constructionism" chapter of *Constructionism* (1991), this does not mean that we should be tempted to describe constructionism by the catchphrase "learning-by-making" (Papert 1991, 1). This is because the "sense of constructionism" with which he is working is "much richer and more multifaceted" (1991, 1) than a one-to-one correlation between knowledge attainment and whatever the individual builds.

Perhaps the most important reason for Papert to assert this concerns a theme that is to be his defining contribution to the constructionist debate. Papert clarifies that the learner actively orientates their construction of knowledge through aspects of the public realm and social consciousness. In learning, the individual engages and manipulates *publicly* identifiable materials (what he calls "entities"), which are physical as well as conceptual. An experiential "sense" of knowledge development is duly established by, and originated through, socially relevant preconditions. This, for Papert, crucially expands upon the constructivist methodology, whereby

> constructionism—the N word as opposed to the V word—shares constructivism's connotation of learning as "building knowledge structures." . . . It then adds the idea that this happens especially felicitously in a context where the learner is consciously engaged in constructing a *public entity*, whether it's a sand castle on the beach or a theory of the universe. (Papert 1991, 1; my emphasis)

I pause here to consider whether, if the learner constructs meaning, is this saying that all knowledge is subjectively produced? Moreover, if this is the case, how is it that common ideas or knowledges are developed socially? How, indeed, would we learn about social time, how would social synchronization occur, if knowledge acquisition is a "reconstruction" rather than a "transmission"? In considering the theme of lateness that occupies this book, if this interpretation of constructionist knowledge acquisition was applied to questions of how one comes to "know social time," the struggle some individuals have in synchronizing with standardized social time structures could be explained. Perhaps their sense of time, as a form of Papert's "meaningful product," would have been "manipulated," that is, "constructed," differently in such instances?

In responding to these concerns, I assert that Papert's sense of the constructed quality of knowledge does not appear to be simply a subjective affair. The reason I say this is that for Papert, the learner's particular reconstruction of meaning has to be contextualized through the social conditions and constitutions of what he describes as "public entities" (Papert 1991, 1). Building on the constructivist notions of cognitive assimilation and accommodation, Papert's constructionism emphasizes that knowledge acquisition is not the transfer of pure, timelessly established, isolated pieces of information to be discovered about these entities. Neither, I would argue therefore, is the knowledge that one has about an entity autonomously partitioned. Rather, knowledge acquisition is a perpetually reconstructive process, both in terms of how one's new knowledges integrate and develop with their existing knowledges, and in how knowledges are conditioned by a *public* realm. This latter factor indicates the relevance of Papert's renavigation of constructivist impressions for an enquiry into collective constructions of time.

In fact, when revisiting the constructivist conditions from which this version of constructionism emerges, we see Piaget describe the child's production of knowledge along reconstructive lines also. The exploration of the world is a creatively interpretative way of being, "there exists invention and no longer only discovery; there is, moreover, representation and no longer only sensorimotor groping" (Piaget 1952, 341). Learning in this way marks a different characterization of a subject's experiential and epistemological relations. This "transition" is from the mode of a straightforward sensory encountering of the world, to that of a constructed epistemology, or in Piaget's terms, from "directed gropings to invention, and from sensorimotor schema to representative schema" (1952, 341). While for Piaget there is a "systematic interdependence" between one's sensory apparatus, and their environmental assimilation and adaptation of the experiential data, it is only through constructive conditions that knowledge production can be traced to one's "representative schemata" (341).

This sense of a representative "invention" that is developed subsequently to the experience of a sensorimotor, natural, physical world (and then acts as an anticipatory filter of it), will generally be in our sights in considering conceptions of the socially symbolic production of time. What the positions of both Piaget and Papert assist us in understanding is a modern line of thought[5] that presents notions of knowledge (re-)construction, and of sequential relations between original conditions and representational schemas. The reading that representational schemas of time-knowledge only manifest after, as invented versions of, a sensory experience with time is implicated in the social constructionist position against which philosophical reflections and interventions will later be appraised.

Despite this consistency with the position under investigation, a critically disruptive potential of Piaget's and Papert's constructionisms could be identifiable in the contestations their theses provide to taken-for-granted assumptions about who owns institutionalized knowledges. This point might accordingly generate a curiosity about the possibly critical elements of constructionism itself. An encyclopedic portrayal of their literal intersection is provided by Milton Campos, for whom "critical constructionism" can be defined as the "merging of constructivist or constructionist views with the critical epistemology developed by the Frankfurt School" (Campos 2009, 216). Critical constructionism is here presented as having an interest in the social conditions of the meanings that individuals construct regarding their actions, when combined with a critical focus on the ethical consequences of the actions that individuals choose consequently (2009, 216). As with Papert's reading of constructionism, for Campos a critical constructionist approach can find beneficial applications in pedagogical theory. This asserts that if certain, constructionist, educational strategies were adopted, they would "'wake up' children's awareness of the underlying logical structures of their own reasoning through critical inquiry" (217). What is notable about the "critical" aspect of this for an educational context is the constructionist intervention that facilitates an individual's self-awareness of—that "waking up" to—the conditions "behind" their present reasoning and decision-making processes. Given that this is also a seemingly *constructivist* ramification, in terms of an individual's cognitive development or "maturation," it is perhaps no surprise that Campos clarifies that "critical constructivism could also be called critical constructionism" (217).

Critical constructionism's typically constructionist themes—encountered in Papert's focus on the role of public entities in knowledge construction—are made more apparent in philosopher Andrew Feenberg's recognition of a critically constructionist drive to communication studies. In *Questioning Technology* (1999), Feenberg states that his work is part of an ongoing critical dialogue that, by recognizing how technological objects could have developed in any number of ways, embodies a "reaction against determinism" that "brought the contingency of technological development into focus" (Feenberg 1999, xvi). Technology is a thematically consistent thread for this stage of our discussion. The socially derived "entities" that are identified by Papert as being crucial to the constructionist perspective on knowledge, are said by him to offer a "learning richness" through embodying processes that are "benefitting from the presence of new technologies" (Papert 1991, 6). Furthermore, considering the time-centric considerations of this book, it must be acknowledged that social time is conditioned by the technologies by which it represents, and dare I say, invents, time.

Feenberg is receptive to characterizations of the continually "invented" or "constructed" path that technological progress takes in society. Citing Marcuse's *Counter-Revolution and Revolt* (1972), as well as Michel Foucault's *Power/Knowledge* (1980), Feenberg reminds the reader that there exists a prominent body of literature that "rejected the idea that there is a single path of progress based on technical rationality and opened a space for philosophical reflection on social control of technological development" (Feenberg 1999, 6). The critical component to Feenberg's examination therefore relates to what we take for granted about present technologies and their apparently natural integration into our lives.

Conversely, it is Feenberg's argument that there are not "natural processes to technological development" (Feenberg 1999, vii), but rather socially, contextually contingent causes and reasons. Feenberg is accordingly comfortable with setting up the natural|social distinction that I contrarily find most curious about conceptions of time, arguing that "to reduce technology to a mere causal function is to miss the results of a generation of social science research. . . . [O]ne cannot reduce technology to natural causality" (1999, 169). As a result of this position, he alternatively declares that the constructivist or constructionist position must critically maintain that "technology is social in much the same way as are institutions" (11). The ramification, according to my reading of Feenberg, of correlating technological development with institutional establishment is the influence that the latter will have regarding the former. Technology manifests politically coercive effects, and institutions come to be technological devices. Rather than technologies marking a straightforward sense of "social progress" or civilizing advancement, specific technological choices bureaucratize or institutionalize society in value-oriented ways. While I earlier countered the reductive natural|social distinction that Feenberg establishes, this latter insight derived from his work can condition pertinent questions about which symbolic technologies are institutionalized regarding time, and indeed, lateness protocols.

It is, however, Feenberg's contention that constructivist and/or constructionist theory does not often venture into such territory. This marks the most critical of his claims, in offering a contrary portrayal of what critical constructionism should be. Just as Horkheimer argues that critical theory is defined by radically interventionist modes, beyond traditional theory's focus on improving what is already in place in the present, Feenberg laments that most constructionist "research has confined itself to the study of the strategic problems of building and winning acceptance for particular devices and systems" (Feenberg 1999, 11). As our discussion ventures more thoroughly into the context of social time, such devices and systems will emerge in the forms of the clocks and calendars that become normalized over long periods.

Constructionist theory becomes *critically* constructionist for Feenberg when it intervenes and disrupts technology systems that have been presently normalized, exposing how technology choices are reflective of established, social, power arrangements. This position becomes most explicit in his "Democratizing Technology" chapter, in which he specifies that "technology is power in modern societies, a greater power in many domains than the political system itself" (Feenberg 1999, 131). A critical appraisal of technological development requires, he demands, that one that is not "ambivalent" (1999, 76) to the power-technology relationship. Offering a primary objective for a critical constructionist approach, Feenberg suggests that such theory must always be identifying and declaring that there seems to be "no unique correlation between technological advance and the distribution of social power" (76). A mode of disciplinary self-reflection is demanded that interrogates how present social hierarchies are often preserved and reproduced through new technologies, as well as via any theoretical endorsements of such reproductive relations.

In examining how the social constructionist time condition exemplifies a sequence of secondariness between natural time and social times, and by considering how philosophies of time exhibit an awareness of this separation in their methods, an attention on how a discipline self-reflects becomes important. The critique of what is taken for granted about social time structures and experiences forces a discipline to review not only traditional positions regarding time but also their own perpetuation of such positions. In returning to Hoy's commentary, traditional theory is distinguishable from this critical mode by taking the perspective that presuppositions about a social structure or theory must be examined while leaving alone the presupposition of its own situatedness in relation to such politics.[6] Conversely, critical theory, Hoy argues, "examines its own place in the social context" via a method that is "self-critical and alert to its own distortions" (Hoy 2009, 228).

Critical theory in this style for Hoy is keenly aware of exploring the development of any particular aspect of the social present, including its own culturally informed awareness of that social present. An acknowledgment of this meta-reflective element seemingly informs Horkheimer's description of critical theory as a "criticism of the present" (Horkheimer 1972, 21), where the present is a reference not simply to an external object of investigation but also to a theory's own germination. This indicates that conversely, in Hoy's reading, traditional theory ignores its own developmental genesis by assuming that how we acquired our beliefs, or knowledge structures, is "irrelevant to the validity of those beliefs" (Hoy 2009, 229). That critical theory attends to what we have otherwise taken for granted about these developments indicates its interrogative mode, "asking precisely how we came to forget the contingency of the historical beginnings of our practices and we persuaded ourselves that these practices were necessary and universal rather than arbitrary and contingent" (2009, 229–30). As indicated, what I later intend to develop from this self-reflective capacity of critical methods is an examination of whether theoretical critique perceives itself to be secondary, or subsequent, to a previously established, present reality. Considerations of the separation of representation from reality, implicit to the social constructionist time condition, are identifiable in such a question.

The terminology that Hoy has just used, suggesting the possible "contingency" of social practices, is notable. During our study of the social constructionist time condition, the interpretation that social structures of time are contingent representations of a separate, naturally celestial time-source is to be comprehensively addressed. Indeed, this could be said to be a potentially emancipatory aspect of a critical inquiry into time, by intending to uncover dogmatic normalizations that marginalize certain demographics from social time's central institutions. This is, of course, a nod to Horkheimer's celebratory description of critical theory, as providing a disruption to a sense of fatalism regarding oppressive social institutions (Horkheimer 1972, 26–27). Time can be seen as one such established, social institution.

In fact, what is apparently entrenched about social time motivates the citation from David Harvey that opens this chapter, whereby time's apparent regularity becomes a social utility, represented by clocks and calendars upon which the social agrees, and to which the social abides. Social theorist David

Gross describes this social agreement as requiring "a particular temporality to be widely accepted and internalized by a population at large" (Gross 1985, 54). I have included Gross's citation here given the interesting application of the word "particular." It suggests that a version of social time currently dominating was not the only way that time could have been represented to humans, or to a particular group/culture of humans, and that even at some point there possibly existed different representations contesting for widespread adherence. In considering socially pluralized constructions of time, it is now necessary to refine the description developed above of constructionist theory by recognizing its specific form of "social" constructionism. Given that this is a study of the *social* constructionist time condition, the *social constructionist* version of constructionist theory is of primary interest. We can review which aspect of social constructionism is most relevant by firstly considering its take on reality generally and then on time specifically.

SOCIALLY CONSTRUCTED REALITIES

As acknowledged of constructionist theory, social constructionism can be defined in multiple ways. While it is not in the specific interest of this study on time to provide an exhaustive analysis of social constructionism, it is appropriate that I explain why I am engaging the specific position of social constructionism that is to be identified.

To preface this explanation, I will provide an advisory regarding the method that I have employed in the following sections. If when moving through snapshots taken from various thinkers you are curious as to why such theory has been chosen, keep in mind that I am attempting to convey a sense of the preponderance of social constructionist tendencies in relatively recent theory. In order to frame the philosophical reflections and destabilizations that are to be offered in later chapters regarding the social constructionist time condition, its existence first has to be emphatically highlighted. This is the process in which we are currently entrenched and will be until chapter 3. By integrating material from a series of theorists, what I am establishing is a continuum of tendencies through which the social constructionist time condition manifests.

Social constructionist theory can loosely be recognized as having a concern with supposed distinctions between human representations of the world and the world that is being represented. These representations can be of a range of objects or contexts, including human categorical divisions such as race, gender, and sexuality, as well as physically material phenomena such as trees, atoms, and whatever else is experienced sensorily. A problem that is raised here regards whether what humans think is "real" about these contexts or objects has actually already been socially conditioned or constructed by the investments of various, normalizing structures or powers. If this is the case, then a social constructionist might argue that the conception and experience of the "real" will be representationally distanced from whatever the "real" is. A gap between representation and reality could mean that a culture's production of scientific physics does not actually grasp the "laws" of the physical environment, or that gender is actually a more complicated affair than the binary mode often culturally adopted.

The inescapability of this constructed experience can be seen as a socially dialectical process, where individuals or social actors in some way create or perpetuate, and are created or perpetuated by, the social environment. This, for Peter Berger, is the "paradox of social existence: That society defines us, but is in turn defined by us" (Berger 1963, 129). One of the first texts to directly characterize this process in terms of the "social constructionist" tag is *The Social Construction of Reality* (1966), authored by Berger with Thomas Luckmann. In setting a platform for subsequent social constructionist theory, Berger and Luckmann conceive that, for social constructionism, what is asked is whether our experience of reality is not attributable to Immanuel Kant's sense of *a priori* structuring frames, nor to a series of psychologically cumulative points of development (Berger and Luckmann 1966, 99, 180). Instead, the suggestion is that an individual's experience of whatever reality is must be socially conditioned.

These themes extend to how social constructionist thought is interested in the ways humans are categorized according to racial, gendered, and sexual parameters, and asks whether such categories are biologically given or socially fabricated. What is being investigated here are *kinds* of humans. Regarding the category of race, for example, a social constructionist theory might posit that our idea of it is either different from its biological truth, or that race actually has no biological truth and thus is a concept that is entirely socially fabricated. Sally Haslanger's "A Social Constructionist Analysis of Race" (2012) captures these scales of thought well, reporting that the most extreme social constructionist theories of race posit a "race eliminativism," in which "talk of races is not better than talk of witches or ghosts" (Haslanger 2012, 299). While this does not necessarily discount race being a natural category, it does argue that whatever might be natural about race has no reality in our socially filtered experience of it, and that in order to achieve racial justice we should just stop participating in any social structure that maintains a sense of race.[7] The critical assertion, intersecting with the constructionist assertion, would here be to ignore race concepts.

Haslanger explains that a less extreme version of this impression of race is found through "race constructionists," who argue that "races are real, but that they are social rather than natural groups" (Haslanger 2012, 299). The point is that race must be recognized not as a natural category but as something that has manifested through the social idea of it. The importance of acknowledging its resulting social reality is traced in the demand that racial justice can be achieved if we "recognize the mechanisms of racial formation so that we can undo their damage" (2012, 299). In these interventionist ambitions designed to emancipate a demographic from a position of structural subordination, we accordingly might again draw a connection between the directions of race social constructionists and the focus reviewed earlier of Frankfurt School critique. A third position, that of "race naturalists," unsurprisingly demands, contrary to race constructionists, that race *is* a natural category, divided on the basis of biological, genetic, and physical features (299). Furthermore, race naturalists ideologically clash with race eliminativists in demanding that recognizing this natural division *is important* in achieving racial justice.[8]

The social constructionist position that separates the natural and social forms of race, while recognizing the reality of each, is the position in which I am most interested. This is because it encapsulates what is most pressing for correlative considerations of a social time that is contingently experientially real but yet manifests as an invented version of time's actually real, natural form. As Ronald Sundstrum anticipates in *Race as a Human Kind* (2002), if we take this constructionist position on race, what must be said is that race *as we know it* does exist. However, the race that we know is based on cultural differentiation, "a socially produced category and identity" (Sundstrum 2002, 91), rather than on actual biological/genetic/physical differentiation. Evoking Berger's earlier encountered characterization of the socially dialectical mechanism of socially constructed phenomena, the constructed categorization of race is argued by Sundstrom to impact on other social categories such as "gender, class, and sexuality" (2002, 91), which in turn reproduce how race is lived communally and individually. An individual's awareness of these classifications will contribute to how they perpetuate the construction of that classification, both via their own identity and for the identities of others. Ian Hacking, who is to become a prominent voice in this chapter, provides an extended discussion on this matter, positing that when demographics and population groups have classifications that are publicly, prominently constructed, "their experiences of themselves are changed by being so classified" (Hacking 1999, 11).

This social constructionist position holds that race has a definite reality in our social lives. As Sundstrom notes though, race becomes real to us because of a structural politics, whereby "race is a *human* kind precisely because it is socio-political, and it is real, because sociopolitics has made it so" (Sundstrom 2002, 101; author's original emphasis). The social constructionist gap between representation and physical reality is duly apparent in the affirmation that according to this perspective, "race is not a real biological kind. . . . [I]t captures no biological truth" (2002, 52). Race constructionism in this model bears the hallmarks of Michel Foucault's conception of sexuality as offered in *The History of Sexuality: Volume 1* (Foucault 1978 [1976]). While Foucault argues that homosexuality is a socially constructed concept, this collectively engendered reality of homosexuality is not discounted. What is required is a recognition that this reality is grounded in the cultural origination of the concept.

Much of what social constructionist theories on race or sexuality in this style posit is that the social categorization of human kinds instantiates the reality of the characteristics associated with the category. This point has been perhaps most notably addressed by John Searle, who in the *Construction of Social Reality* (1995) argues that "the attitude that we take toward the phenomenon is partly constitutive of the phenomenon" (Searle 1995, 33). "Social facts" are distinguishable in this regard from "physical facts" for Searle, in that it is suggested that the discursive production of social facts produces a kind of reality that is not equally possible of the latter. For instance, a collective attitude that recognizes homosexuality will produce, it is argued, the fact of homosexual identities in a society. However, the same kind of generation cannot be said of physical facts. Just because a collective population believes there is no gravity, this will not mean that the physical world is devoid of gravity, marking for Searle "a remarkable feature of social facts" and one that "has no analogue among physical facts" (1995, 34). We see the earlier constructivist hallmarks reemerging here, in which social truths are "invented," while sensorimotor truths are dependent on "groping" and "discovery."

Perhaps the most straightforward example of a socially constructed fact, or of what Searle describes as a "social fact," is presented in Stephen Pinker's *The Blank Slate: The Modern Denial of Human Nature* (2002). Pinker notes that certain categories of fact are social constructions simply because their actuality is entirely due to human, social consensus, and "bear no systematic relation to things in the world" (Pinker 2002, 202). With examples such as money, citizenship, or the presidency of the United States, "they only exist because people tacitly agree to act as if they exist" (2002, 202).

For Searle though, this is somewhat of an oversimplification. Searle maintains that every social fact will ultimately actually depend on an antecedent physical reality, or what he describes as a "brute fact" (Searle 1995, 190). In explaining the conditions for a socially constructed reality that are found in brute, physical reality, Searle wants to show that even socialized speech acts about matters that seem to be entirely culturally originated "refer to a reality beyond themselves . . . a reality independent of all representation" (1995, 190). Given that my interrogation of the social constructionist time condition is interested in suppositions of a temporal reality that is chronologically and ontologically independent from the socially constructed versions of it, this separation of physical reality and social representation is the kind of reading of the social constructionist position in which I am interested.

Searle exhibits this supposition by arguing that when we construct money, property, and language, for example, each is conditioned by the "raw materials of bits of metal, paper, land, sounds, and marks" (Searle 1995, 190). Money is produced from ink and pulp-based materials that are then constructed socially for a system of market exchange. At the extreme point of this assertion regarding the independent, brute conditions, of all social realities, Searle notes that when we conceptualize a mountain, the mountain itself is a mountain "even if no one believes it is a mountain" (1995, 33). If one was to deploy Searle's terminology to the stakes of an investigation regarding time, these "brute physiological phenomena independent of all representations" (191) would be the rhythms of celestial, physical, universal bodies. A brute rhythm is then duly positioned as the inaccessible base of social forms and ideas of time.

As a result of this "logical dependency" of socially constructed facts on brute, physical facts, Searle argues against the notion of a universal, social constructionism that posits that everything is socially constructed. The reason for Searle's contestation to the idea that all facts are social or institutional is that, as the previous interpretation reveals, social facts "are logically dependent on brute facts" (Searle 1995, 56). A social reality lacking a physical reality upon which it was representationally based would "produce an infinite regress or circularity in the account of institutional facts" (1995, 56), with no underpinning, originary, or real ground. I posit accordingly that suppositions concerning the logical dependence of social phenomena on the physical phenomena that they represent capture the aspects of social constructionism that suggest there are gaps between the "real world" and the way that "real world" is represented. It is this form of social constructionist inquiry that speaks most directly to my concerns regarding distinctions between natural time and social time.

It is worth noting that the theoretical reading of social constructionism to be used most prominently for my formulation and interrogation of the "social constructionist time condition" is that presented by Ian Hacking in *The Social Construction of What?* An important element of Hacking's presentation is how, having researched a wide range of articles and books that feature the words "social construction" in their titles and content, he provides a comprehensive snapshot of approaches considered to be "social constructionist." Within Hacking's commentary can be found the already established characterization of social constructionism as holding that our experience of the world is filtered by representational versions. Moreover, what Hacking illustrates is that a common social constructionist position is that one is unable to *directly* experience a pre-social world, or indeed a world outside our social habitation. We instead live via the representations of that world, whereby "a great deal (or all) of our lived experience, and of the world we inhabit, is to be conceived of as socially constructed" (Hacking 1999, 6). This "inhabited filter" means that our knowledge of the pre-social world must be questioned.

Hacking goes on to explain the contestation that social constructionist perspectives make to this notion of "the given," or natural inevitability. It is proposed that this challenge occurs via the argument that any particular aspect of society, "X," that is seemingly naturalized, "need not have existed, or need not be at all as it is. X, or X as it is at present, is not determined by the nature of things; it is not inevitable" (Hacking 1999, 6). What is installed through this discussion of inevitability, as I read it, is a conceptual distinction between nature's inbuilt necessity, and social contingencies. The presumed differentiation of nature's singular order, from the myriad directions of social fabrication, is integral to the impulse that brings me to this inquiry into time.

A critique that Hacking offers of this aspect of social constructionist thought is that often it is not clear, or is "fuzzy at the edges" (Hacking 1999, 21), regarding what is not "inevitable." Hacking's response to this issue is to look at social constructionist literature that is concerned with scientific conceptions of phenomena to illustrate that what is typically argued to be "constructed" is the *idea* of the thing, rather than the thing itself. In drawing upon Andrew Pickering's *Constructing Quarks: A Sociological History of Particle Physics* (1984), Hacking posits that "Pickering does not claim that quarks, the objects, are constructed. So the *idea* of quarks, rather than quarks, might be constructed" (Hacking 1999, 68; author's original emphasis). Pickering does not deny the reality of quarks. Instead, what is forwarded is the contingency of our conception of quarks, whereby according to Hacking, Pickering "maintains only that physics did not have to take a quarky route" (1999, 70). For Hacking therefore, social constructionism in this form accepts that an objective reality exists, and is concerned with the ways that detached social knowledges about such reality are produced. Similarly, I interpret regarding the social construction of knowledge about time, that there is the belief that the celestial, planetary, temporality of the universe exists, indeed to the extent that it is positioned as the condition of the social versions we have of it.

The debate concerning Pickering's work exemplifies an aspect of the social constructionist problematic that asks how "true" or "objective" a culture's scientific measurements and resulting symbolizations of the universe are perceived to be. As Hacking notes, our schooling presents the idea that science provides a unique point of access to reality, in that "the history of science tells of definite bench marks, established facts, discovered objects, secure laws, on the basis of which subsequent inquiry proceeds" (Hacking 1999, 32). However, if a social constructionist argument is that all socially developed institutions could have developed in myriad ways, then scientific experimentation, scholarship, and laws themselves, could also have. The set of "natural laws" that science presents regarding the world was not inevitable. Rather, in tracing how they have developed, our perspective needs to turn to institutionalized (perhaps economic or political) forces.

Around the same era as Pickering's work, the developing fields of sociology of science, and science and technology studies, were employing what can be described as social constructionist techniques to interrogate the "objective status" of the facts garnered and processes developed by the physical sciences. This was designed to illuminate the roles of both human subjectivity and cultural normativity in the enactments of scientific practice. Bruno Latour's pioneering work in this field is well known and is illustrated in *Laboratory Life: The Social Construction of Scientific Facts* (1979). Written with Steve Woolgar, the argument is made that there are mediating forces of language and writing, as well as of the constitutive technologies of the laboratory, that shape the "scientific facts" derived. The laboratory is not seen as a neutral, fact-gathering zone but as a socially fabricated and sociologically analyzable arena in which the sociality of science becomes apparent. Pickering's study would later represent a similar kind of investigation, arguing not only how socially contingent aspects of scientific truth were, but also how a scientist's subjective value orientations and the "idiosyncrasies of personal biography" (Pickering 1984, 246) come to inform the premises of a scientific theory and the construction of experimental apparatuses.[9]

As these kinds of readings of the physical sciences gained notoriety, so the inevitable backlash from scientists would become more prominent. Known as the "science wars," this series of intellectual exchanges about the relationship between scientific realism, and the postmodern critique of it, came to a dramatic head in the 1990s.[10] Briefly running through key interchanges in this debate will assist us shortly in further situating the position on social constructionism that we are interrogating.

Early in the 1990s, philosopher Richard Boyd frames the "naturalist's reaction" to constructivist challenges to scientific authority by observing an "incompatibility" of "logical empiricism and scientific realism" with "constructivism" (Boyd 1992, 133). This is largely attributable to assumptions of a singular knowledge or truth being possible in the former, opposed from the interpretation of a cultural pluralism seemingly defining the latter. Taking the relations between the two fields to a seriously acrimonious level, Paul Gross and Norman Levitt's *Higher Superstition: The Academic Left and Its Quarrels with Science* (1994) would then come to characterize the "peculiarly troubled relationship between the natural sciences and a large and influential segment of the American academic community . . . we call here 'the academic left'" (Gross and Levitt 1994, 2). Gross and Levitt posit that the "academic left," a grouping comprising constructionists, cultural theorists, and "postmodernists," lacks a defined theoretical position apart from being "unambiguously hostile" to science (1994, 2). The authors detail, and take great offense at, the constructionist's idea that "society's objective science is thus to be read, in large part, as a construction of its ideological commitments" (64). For certain commentators of the period however such as Andrew Ross, the *influence* of such responses illustrates how socially produced our conceptions of knowledge and truth always inescapably are. As Ross states in *The Science Wars* (1996), the "new science studies show how the 'order of knowledge' has also been the 'order of society'" (Ross 1996, 16).

The disjunction between realist and constructionist interpretations of the socialization of scientific representation reaches a crescendo when, in 1996, physicist Alan Sokal submits an article to an issue of the cultural studies journal *Social Text* dedicated to the science wars. The article is written to covertly mimic the constructionist vernacular found in the journal. Describing how "postructuralist critiques have demystified the substantive content of mainstream Western scientific practice, revealing the ideology of domination concealed behind the façade of objectivity" (Sokal 1996a, 217), Sokal's ruse is instead actually designed to show, in his opinion, how speculative and unsubstantiated the social constructionist's position regarding science is.

It is through engaging Sokal's associated accusations that in the next two paragraphs I can articulate a reminder of the specific social constructionist position in which I am interested. Sokal disingenuously follows on from the above claim by asserting that as a result of constructionist insights, "physical 'reality' is at bottom a social and linguistic construct" (Sokal 1996a, 217). While Sokal does not believe it, in one regard this portrayal is apparently valid, in that we do unavoidably live or experience everything, including physical reality, through socially conceived prisms. However, I want to highlight that, as explained via Hacking's reading of Pickering, for social constructionism this does not necessarily mean that physical reality, the quark itself, is socially constructed. Instead, our *idea*, our representation, of such reality, is what is constructed. Social constructionism in this mode installs an unassailable gap between a physical reality that exists in itself—unconstructed—and our socially constructed experience of it.

One reason I have incorporated this debate into my inquiry is to illustrate a common ground that conditions, ironically, both sides of the science wars. By noting these common conditions, I can further clarify the angle of our interrogation of the natural time, social time, problematic. Regarding these common conditions, what I interpret is that both the scientific realists, and the social constructionist position that I have targeted, install a belief in the existence of a natural, real domain. While each side holds opposing views regarding our possible access to what is naturally real, both suppose that there is a realm of natural phenomena that socialized humans symbolically represent. One such phenomenon, of course, is time.

If we review other sources of Sokal's thought, we in fact find that he offers readings of social constructionism that are consistent with the position within it that I have just presented. Immediately following the appearance of the *Social Text* article, Sokal publishes a short piece in *Lingua Franca*, "A Physicist Experiments with Cultural Studies," in which he reveals his intentions behind his work in *Social Text*. These intentions are declared as relating to his incredulity that the social constructionist position is either constituted by a "dogma . . . that there exists no external world. . . . Or that, there exists an external world but science obtains no knowledge of it" (Sokal 1996b, 61). Consistent with my critique in the previous passage, Sokal's second characterization of the social constructionist interpretation, of an external world that exists outside the socially constructed, scientific conception of it, is the position with which our inquiry into time is engaged. Indeed, this position is inherent to what I have identified as the "social constructionist time condition," that being the *separation* between a naturally celestial time and social times, rather than the denial of either component.

My underlying point here is that the separation of the socially constructed phenomenon, from the "real world" phenomenon that it represents, is evident in those who champion, and those who criticize, constructionist methodologies. Philosopher Paul Boghossian provides one such prominent voice a decade on from Sokal against the cultural relativism that a belief in this representation|reality gap might facilitate. For Boghossian, if the objective, physical world is inaccessibly separate from the various social representations of it, then an extreme constructionist interpretation could posit that all such representations are as valid as each other. If I put this in terms of the question of time, this would require the perception that one culture's symbolic representation of time captures the truth of the physical, celestial, natural time, "as accurately" as any other culture's representation does.

For Boghossian though, instead of subscribing to the belief that any view of the world is as true as any other, we have no "choice but to recognize that there must be some objective, mind-independent facts" (Boghossian 2006, 57). The reason for this is an extension of Thomas Nagel's famous argument that if "everything is subjective," then that claim itself must be taken objectively, which simultaneously contradicts the claim's central condition (Nagel 1997, 15). Boghossian's response to the possibility of constructionist relativism is that in saying "there are no absolute facts" but only "relative facts," what the constructionist is actually arguing is that there actually *are* absolute facts implicated in conditioning the belief in relative facts. As an example, Boghossian notes that the constructionist will not concede that humans have "objective facts" or knowledge about the world. Instead, they will only agree that we construct our knowledge of the world "according to a theory that we accept" (Boghossian 2006, 54). Rather than a global relativism, however, what this actually posits for Boghossian is a belief in the absolute facts "about which theories we accept" (2006, 54). Or in other words, there are objectively factual markers apparent in constructionist ideologies.

I would like to suggest that this reading of an implicit absolute that lurks in the constructionist model can be identified in what we could call the "interventionist stage" of the social constructionist method as conceived by Hacking. This requires returning to his argument that for the social constructionist, when appraising a structural present, X, its contingent construction is often characterized as that which has subjugated a certain population. For Hacking, the social constructionist accordingly argues that not only is X a contingent, social development (and therefore changeable) but also that "X is quite bad" (Hacking 1999, 6). Given these negative ramifications of constructed X, Hacking posits that social constructionist theory takes on an interventionist responsibility by suggesting how we would be *absolutely* "better off if X were done away with, or at least radically transformed" (1999, 6).

In terms of the politics between what is perceived to be natural versus social, the implication above is that while the natural order of things has a determinacy with which we seemingly cannot easily interfere, the contingent composition of any socially constructed frame, as well as our awareness of its socially constructed constitution, can shift. This is somewhat affirmed by Boghossian's critique of social constructionism. In asking what is the point of exposing construction wherever it exists, for Boghossian "a social construction claim is interesting only insofar as it purports to expose construction where none had been suspected, where something constitutively social had come to masquerade as natural" (Boghossian 2006, 18). According to this definition, for social constructionism to be *critically* useful or interesting, it would not only have to illustrate a reconfiguration of a present structural oppression but also be required to "uncover" a situation where the social convention was assumed to be the consequence of a naturally deterministic inevitability.

What has been established, therefore, is a reading of social constructionism in which both natural and social realities are acknowledged but where only the social is directly accessed or experienced. This interpretation that our sense of, or feel for, a natural, material reality, is a socially constructed contingency means that for certain social constructionist perspectives the divide between natural and social realities is insurmountable. Adrian Bardon provides one such commentary, noting in *A Brief History of the Philosophy of Time* (2013) that a socially constructed condition means "we can never penetrate to the sheer naked reality of things as they are in themselves, unmediated by the conditions under which we experience things" (Bardon 2013, 176). The lack of inevitability about our socially mediated, institutionalized, impressions of reality is framed here as a corrupting or warping lens, for "whatever we come up with as a description of nature will always represent a particular way of understanding nature and never a final, unique, fully independent description" (2013, 176). Nature and culture must duly be oppositional realms, with Bardon lamenting that "there is no way for us to step outside ourselves as a species and directly compare our representation of nature with nature itself, in order to see if the former is an accurate reflection of the latter" (176). As I interpret it, the inaccessibility of nature means not only that the representation of it can only ever be an approximation, but that we will never really know how *mis*representationally we live.

What should be clear at this point is how presumptions of this sequence between natural and social phenomena inform my interest in identifying a condition within social constructionist positions that separates natural time and social times. I have framed the impending analysis of this "social constructionist time condition" through an extended discussion with relevant appraisals of the general elements of social constructionist perspectives. Now we shift our focus to impressions of the social construction of time specifically.

SOCIALLY CONSTRUCTED TIMES

If according to social constructionism our experiential reality is defined by contingent developments and parameters, then time must be experienced in the same way. Exhibiting the methodological self-awareness earlier demanded by critical theory, Adrian Bardon notes that "before we can formulate questions about time, we need to look carefully at what our notions about time include, and what facts and concepts we take for granted in both colloquial and scientific discourse" (Bardon 2013, 3). For Bardon, in an everyday regard, what we presume, or take for granted, to be true about time is that its socially symbolic measurement is time itself. This concerns Bardon, who demands that this "doesn't tell us much" about time's actuality, because what is being measured is duration, and duration is a constructed temporal concept (2013, 3).

I characterize Bardon's point as constructionist because in analyzing the measurement of time, we are said to be not accessing the reality of time but rather symbolic representations of its relative lengths (durations). The social constructionist polarization of natural and social phenomena emerges in this account by installing an ontological gap between the celestial, universal reality of time, and the culturally constructed forms of it. Time is most susceptible to a constructionist error according to Bardon, who states that "our grasp of time will always be mediated by our way of understanding things. Temporal experience is a kind of construction, rather than a mere reflection of nature" (Bardon 2013, 176). The ramification here is that as soon as measurement occurs, representational distance entails. This anticipates a key aspect of Bergson's philosophy of time that is to be engaged later in this book.

One of the points inherent to Bardon's commentary is that socially constructed time symbols are ingrained in our experiences of the world from the moment we are born. We can find this sentiment articulated in other constructionism-infused commentaries about time. Michael Flaherty notes in "The Perception of Time and Situated Engrossment" (1991), that as various forms of philosophy, social theory, and sociology explore the historical and structural organization of time, they "illuminate the social construction of time as it is manifest in clocks, calendars, and schedules" (Flaherty 1991, 76). These constructed objects of time represent something tenuous about an individual's temporal experience for Flaherty, in that one's consciousness of them becomes erroneously ingrained as being of "time itself." Such objects are not exclusively the industrial objects found in clocks and calendars but can also be ideological objects. Anthony Aveni observes in *Empires of Time* (1989) that time in this regard can only be appreciated by the collective construction of "the images and metaphors" that are used to "conceptualize time" (Aveni 1989, 8). The centralization of the clock in these consciousnesses of temporality is a common theme in accounts that demand that while we take for granted the reality of clock time, there is a natural temporality underpinning it that is often neglected and out of reach. Kevin Birth explores such accounts in *Objects of Time: How Things Shape Temporality* (2012), reporting the interpretation regarding a life lived according to socially constructed time-symbols, that the temporalizing objects that house them "have an artifactual existence that mediates between consciousness and the world" (Birth 2012, 9).

What matters about these perspectives is that they begin to evidence the interpretation of an oppositional distinction between naturally celestial time and socially structured times. This separation is a key aspect of social constructionist reflections on time. Birth's appraisal of the way that time is socially constructed notes that when a knowledge of the date is required, "one consults a calendar rather than observes the Sun, Moon, and stars" (Birth 2012, 9). Discourses such as these equate human time-knowledge exclusively with cultural production, distinguished from the natural phenomenon by which the cultural production is apparently conditioned. Similarly, in *Time and Social Theory: Towards a Social Theory of Time* (1992), Helga Nowotny commentates on a history of work in philosophy, social theory, sociology, and anthropology in which the formation of time concepts, and the making of time measurements, are exclusively issues regarding social knowledges, whereby "knowledge about time is not knowledge about an invariant part or objects of nature" (Nowotny 1992, 436). Socially constructed frames of time-knowledge are rather apparent in the way time-knowledge is transferred between generations, "inherent in the societal evolutionary process" (1992, 436–37). According to Nowotny, it is via the intergenerational transfer of time-knowledge that this socially evolutionary process is seen to emphasize a progressively expanding distinction of the socialized/culturalized phenomenon from its natural roots, and be steered towards humanity's "gradual process of mastering the external natural environment" (437).

The sense of a civilizing, socializing, and mastering of a natural temporality can be linked to an increasingly regimented form of social time. For historian Daniel Boorstin, social time symbols represent the supposition of humanity's time-centric escape from nature, where only by "marking off months, weeks, and years, days and hours, minutes and seconds, would mankind be liberated from the cyclical monotony of nature" (Boorstin 1985, 1). A socialized control of the otherwise unwieldy, natural phenomenon of time, hence comes to indicate how time is afforded genuine meaning. Simonetta Tabboni's contributions to this debate are important given her recognition that as a culture observes material, spatial transition, it recognizes nature's cyclical changes and then representationally assigns a meaning and direction to such change (Tabboni 2001, 7). By "conferring meaning" to universal change through social symbolization, what was previously stuck in a natural cycle of identically repeated transitions is invested with progression through social rewiring. As Tabboni states, the notion that time is only experienced through socially structured parameters reconfigures the value of "the regular cyclical return of the same phenomena," which without social stratification actually "seems unchanging" (2001, 8). Equally for Barbara Adam, social time is characterized as responsible for the "structuring of 'undifferentiated change' into episodes," establishing "natural time as very different from its social science conceptualisation" (Adam 1990, 150–51).

The changing dynamics of social time present, according to Hacking's earlier encountered portrayal of social constructionist theses, as a hallmark of social phenomena. Both within the same culture, and across different cultures, social time's capacity to change is reflective of its already contingent, or contextual, construction. Tabboni agrees:

> Human societies construct changeable ways of measuring time with the non-changeable purpose of connecting change to the meaning they intend to confer on collective works, history, and changeable life in general. (Tabboni 2001, 9)

A critical attention can be directed toward this insight that socially structured time is a changeable measure rather than is naturally fixed and determined. This is because, in cultures where the symbols and structures for social time do not change over many generations, there could be something about the normalization of those structures that benefits certain institutions and populations while being detrimental to others. It is in such circumstances that for Tabboni, the "acceptance of a temporal norm has taken a form which, having passed through a long series of historical transformations . . . has become continuous, uniform . . . demanding and persuasive" (Tabboni 2001, 10). How theories of time critically appreciate this continuous force of normalized time and its historical development (2001, 12), is for Tabboni one of their central mandates.

The suggestion therefore is that the continuous force of normalized, social time in any particular cultural context can be distinguished from social time's globally differentiated forms. The week, for example, has a history of varying in length between different societies. Historians Peter Coveney and Roger Highfield observe that this arose from societies needing "a time unit smaller than a month but longer than a day" (Coveney and Highfield 1990, 43). The ancient Colombians had a three-day week, the ancient Greeks ten-day weeks, whereas the contemporary, Western, seven-day week is a Babylonian legacy that influenced the Jewish calendar, whose week consisted of six days plus the Sabbath.[11] The French decimalized the Judeo-Christian seven-day week in 1799. However, as historian Gerald Whitrow observes, "their ten-day week was soon scrapped by Napoleon" (Whitrow 1989, 120). More recently, the Soviet Union fluctuated between five-day and six-day weeks.

In analyzing the social constructionist time condition, what I will be asking in this book is whether social time structures are mere changeable versions of what time actually, naturally is. This question matters not only for readers with a philosophical background who are interested in the interdisciplinary interventions that the pivotal philosophies in this work might exhibit, but also for all readers for whom the distinction of a *real* time from the socialized times in which we live should engender an interest. After all, one's socialized time experiences feel real, rather than merely (mis)representational forms of what time supposedly really is.

As we will later see, conversely for Émile Durkheim, time is only a social phenomenon, meaning *human* social time. This interpretation permeates the introduction of *The Elementary Forms of Religious Life* (1915 [1912]), in which Durkheim declares,

> the category of time expresses a time common to the group, a social time, so to speak. This category itself is a true social institution. It is peculiar to man; animals have no representation of this kind. (Durkheim 1915, 11)

It is when time is characterized as an exemplar of human exceptionalism in this manner that we see conceptions of its socially constructed state readily correlated with modern, industrial, and commercial developments. The sociology of Georg Simmel becomes relevant in this regard, where he explores whether the individual in the modern city can "preserve the autonomy and individuality of his existence in the face of overwhelming social forces" (Simmel 1997 [1903], 174–75). One such social force is the common, imposed time by which everyone abides in order to socially function. Life within this new, thriving, metropolitan environment requires coordination given the central role of the metropolis as the "seat of the money economy" (1997, 176). The city can only manage this responsibility by arranging its subjects objectively. Metropolitan life duly integrates all activities into a reliable and impersonal time schedule to which citizens conform. Here Simmel employs the common interpretation of time as a mechanism from which the idiosyncrasies of individuals are excluded, observing "a firmly fixed framework of time that transcends all subjective elements" (Simmel 1971 [1903], 328). Subjects adhere to this structural objectivity in order that the particularities of their lives cohere with the social collective.

Such accounts inform certain characteristics of enquiries into time in which the focus is the social management, or experience, of time.[12] In this guise, earlier identified in the work of Michael Flaherty, an objectively separate source of time is presumed—a natural, worldly rhythm—anterior to the social abstraction and organization, or the subject's experience, of it. By interrogating this supposed separation, which exemplifies the social constructionist time condition, I will be interested in how Derridian, Merleau-Pontian, and Bergsonian philosophies reflect on notions of time's supposed "abstraction." Barbara Adam is aware of these kinds of objective|subjective distinctions often found in sociological theory, stating in *Time and Social Theory* (1990),

> To study the experience of duration, the estimation of an interval, people's orientation within horizons, or the timing, sequencing, and coordinating of behaviour, is to *define time* as duration, interval, passage, horizon, sequencing, and timing . . . time does not "emerge" from these studies, but is pre-defined in the very aspects that are being studied. (Adam 1990, 94; author's original emphasis)

What is being stated here is that in discussing time, humans merely refer to the social category they have already preconstructed. This attributes time-reckoning capacities to humans, which Adam posits is identifiable in how the "members of Western industrial societies . . . *create time* as a resource, as a tool, and as an abstract exchange value" (Adam 1990, 161; author's original emphasis). However, I would like to add that in arguing that humans create time as an abstract "resource" or a "tool," Adam is seemingly endorsing the reading that the time of which humans are conscious is simply a socially meaningful or manipulable abstraction of a more fundamental temporality. Indeed as is indicated above for Adam, time does not "emerge" from the socially sanctioned study of it whatsoever.

The relative strength of Adam's work in my view is her intention to critically analyze time's "pre-defined" givenness, exemplified in the hope that "time will lose some of its taken-for-granted status in social theory" (Adam 1990, 68). Adam's commentary in this context illuminates pertinent aspects of social life, providing a comprehensive default reference for any researcher of social time. In *Timewatch* (1995), Adam states that human, social behavior does not simply arrange time but in some way constitutes time. One such observation notes that "human social life . . . constitutes time, entails time and is enacted in time: it creates a new past and a new future" (Adam 1995, 39). My concern from this is that the consequent critique of what social theories of time have taken for granted about time avoids addressing the social constructionist supposition or condition of a separately anterior, temporal reality that is being misrepresented. This is not necessarily a criticism of Adam's legacy but an indication of how her concerns might be distinguishable from mine.

This point further differentiates the direction of my research from the accounts of time provided in the renowned work of Michael Flaherty. Flaherty generally explores the subject's *experience* of time. This is declared in *A Watched Pot: How We Experience Time* (1999), as a response to the tendency of "students of temporality to gloss over the basic fact that what feels like minutes for one person may feel like hours for another" (Flaherty 1999, 5). By interviewing subjects in different social arenas, what emerges is that the experience of time is not universally identical. Consequently for Flaherty, in particular circumstances, it "*feels like* much less time has elapsed than has actually been measured by the clock or calendar" (1999, 104; my emphasis).

As has been indicated, the purpose of integrating empirically engendered observations such as Flaherty's is not to frame an exploration of our own through the diverse abundance of subjective temporal experiences. The intention instead with the incorporation of such material is to note the existence of empirical studies that posit a distinction between time as it is represented idiosyncratically or contingently, and a separate actuality of time which conditions such representations. Analyses such as Flaherty's seemingly presume the existence of an anterior, objective time-source that grounds the differing experiences/impressions of it. Rather than adhering to such a sequence, in this book we will be considering how philosophy interrogates the conception of an inaccessible time-source that governs contingently felt or structured, separate experiences of it. Philosopher Gail Weiss expresses similar reservations about subscribing to a model that conceptually separates the "actual" of time from human experiences of that time. As Weiss emphasizes, "[S]urely it is overly simplistic to say that time, as measured by calendars, watches, sundials, and the movement of planets and stars, is 'out there' while our temporal experience is within us" (Weiss 1999, 112). For the social constructionist position being engaged, the "within us" of which Weiss speaks marks a socially engendered internality. From such collective conditions are derived our subjectively "internal" impressions. In distinguishing these impressions from "actual time," terminology such as "feels like" and "seems" is peppered throughout Flaherty's work, as sensations that are separate from the real, "out there" temporality that is measured by clocks and calendars (Flaherty 1999, 29, 34, 37, 59, 64, 94, 104). As with my commentary on Adam, I raise this not as a straightforward criticism; rather it flags the differentiation of Flaherty's research focus from mine. We will later return to the social intersection of time-experience and subjectivity in Flaherty's work.

My interests in this book represent a selection of the many ways to approach the question of how time has been constructed according to social and subjective requirements. In recognizing this, it seems appropriate to acknowledge how eloquently this theme of differing research intentions is handled by Austrian philosopher and social theorist Alfred Schütz, whose phenomenological project asks if theory that attends to social phenomena is "concerned with the *very being* of man or only with his different modes of social behavior?" (Schütz 1967 [1932], 3; my emphasis). While my focus is on that "very being," the different modes of social behavior are not opposed to, nor alienated from, such an inquiry. Indeed, I believe such modes can be intentionally engaged in order to leverage insights into the "very being" of Being generally. One such mode, one such behavior, one such socially regulated structure by which we can appreciate the conditioning suppositions implicit to the social construction of time, and to social constructionist theories about the social construction of time, is lateness.

Chapter Two

Relatively Late

Cultural Plurality and Modified Bodies

All things arranged temporally are taken from social life.
—Émile Durkheim, *The Elementary Forms of Religious Life* (1995 [1912], 10)

THE GLOBALLY RELATIVE REGULATION OF LATENESS

In order to interrogate the social constructionist time condition, in which the myriad forms of social time are positioned as secondarily separate to a singular, natural time, a lens is required through which the relative plurality and variability of social time is evidenced. This lens, for this project, is lateness. As explained in the introduction, there are four contexts through which lateness is examined in this book. All have been chosen because of how they embody the secondariness by which socialized temporalities are conceptualized in relation to a natural foundation. Those four contexts are: (a) protocols regarding social synchronization, (b) socially modified and constructed bodies, (c) the cultural production of theory, and (d) responses to climatic and ecological change. I expect that all readers would be familiar with concerns of being socially "on time," as well as live in societies in which body modification practices occur. Likewise, this book is being written in an era in which a collective consciousness regarding the environmental effects of human cultures and development is expanding. These case studies are duly designed to give this philosophical inquiry a general relatability, or at least to provide a set of contexts through which a reader might feel invested in my concerns about the status of constructed time(s).

In engaging the first context, it can be noted that as sociological and anthropological theory attends to the various ways that time is constructed and experienced by cultures around the world, what emerges are differing social regulations for being late. The discussion that is to follow incorporates characterizations of time from the Philippines, the Caribbean, a conglomerated sense of Africa, and an equally conglomerated sense of "The West." In engaging theory that collectively discusses "Caribbean time," or "African time," I do not suggest that all the countries and cultures of the Caribbean, or of Africa, respectively, share identical outlooks on time. Furthermore, when integrating theory about "Western time," neither do I wish to propagate a fallacy that *all* Western time structures and rules regarding lateness are similar, in opposition to all those that are somehow equally non-Western.

Rather, what I will be looking to appreciate via commentaries on regional differences is the belief in a global plurality or multiplicity of socialized time. Of further interest is what these regional differences are claimed to tell us about a gap or distinction that is posited between natural time and social times. As we unsettle this distinction, readers might find themselves feeling increasingly involved in the aspects of this inquiry that engender broader reflections on which cultures are portrayed as being less developed than other cultures, and why understandings of time inform such portrayals. Be warned, this will be an extended whirlwind of a discussion, spanning this and the next section within this chapter, crossing many spatial and temporal boundaries, and incorporating as many voices as is briefly yet relevantly possible. Once the form of the social constructionist time condition, separating natural time from social times, has been established in these early chapters, we will then be able to appraise how the philosophies of time already indicated critically reflect on such a position.

By dispersing the attention on social time among different zones of the world, I hope to avoid the problems that "chronocentric" models exhibit. The term "chronocentrism" was conceived by sociologist Jib Fowles to refer to the belief that the time structure in which one lives is the default (Fowles 1974, 65). The acknowledgment that is soon to be presented here, of the plurality of socialized times, potentially counters the embedded persuasiveness of chronocentric assumptions. This is not just my impression. Integrating such considerations into time-theory encourages, according to scholars including Ulfried Reichardt (2000) and Helga Nowotny (1975, 1992), European-centric perspectives in particular to appreciate the contingent, rather than the default, status of the time structure in which one is immersed. In order to avoid chronocentric reductions, for Nowotny, one must "take into account the plurality of social times" (1975, 326).

An appreciation of the plurality of social time can, I believe, be facilitated by an attention on different, culturally localized, regulations of social synchronization and lateness. In considering the time conventions operating in the Philippines, philosopher Darren Gustafson exemplifies a tradition of distinguishing between cultures that adhere to the industrial mechanization and symbolization of chronological time versus those that do not. Lateness in the Philippines is consequently described as operating according to less strict protocols than in cultures or contexts where the twenty-four-hour clock is centralized. In arguing that "'Filipino time' was never chronological to begin with" (Gustafson 2012, 90), Gustafson makes the point that if we correlate chronological time with an industrialized, Western perspective, then how we view non- or less-chronologically inclined social time structures needs to appreciate the situated politics involved in cross-cultural, temporal comparison. Nick Joaquin picks up on this point in analyzing the way in which many Asian temporalities are reviewed by Western cultural theorists. Joaquin reiterates Gustafson's point of Filipino time lacking chronological origins, describing this as "a quality lingering over from the 'timelessness' of an old culture, and as a dogged resistance against the advent of the foreign-tyrant clock" (Joaquin 1982, 164–65). Filipino time is accordingly portrayed as qualitatively differentiated from industrial, Western standards. From this perspective, accounts like that of social theorist Heather Gingrich emerge that make the rather broad claim that in a country such as Canada, "high value is placed on being on time for appointments" (Gingrich 2006, 13). Conversely in a country such as the Philippines, Gingrich posits that "time is much more flexible" (2006, 13). I am not insensitive to the broad generalizations being made here concerning the reduction of cultures to certain tendencies regarding lateness protocols. To the contrary, it is through being attentive to, and acknowledging, these kinds of characterizations that we can develop a sense of what our earlier engagement with constructionist theory describes as the contingent constitution of social phenomena.

That clocks do exist and are used in the Philippines prompts queries about the reasons behind a relative frequency of late behavior. The influential work on American-Philippine relations by Joseph McCallus notes in this regard that American perspectives on Filipino temporality question why tardiness is so instilled in a culture that does have clocks and wishes to progress into the "developed world." These kinds of inquiries typically venture to ask, "[A]re Filipinos always late by tradition or because of laziness" (McCallus 1994, 51). The theme of the plurality of social times here emerges, with McCallus observing that "if ever there was an unsuccessful case of imposing Western . . . culture standards on Filipino culture, it is the stubborn refusal to follow 'objective' time'" (1994, 51).

Larry Purnell and Betty Paulanka identify correlative, theoretical assumptions regarding timeliness being of greater importance to "North Americans," when drawing comparisons with "Central and Southern Americans." While it is argued that North "Americans are supposedly expected to be punctual on their job," with an "apology expected if one is late by more than 5 or 10 minutes" (Purnell and Paulanka 2003, 21), this is said to not be the case in certain other global contexts. Their work cites research undertaken in Panama, evidencing that a meeting starts not when the clock dictates, but whenever the "majority of the people arrive" (2003, 21). This is attributable, it is explained, to the assumption that Central Americans share the African American trait of being more "present- than past- or future-oriented." The claim is that a present-oriented perspective places greater importance on the "here and now," than on what "may occur in the future or has occurred in the past" (17). Accordingly, if one is not as concerned about what is happening beyond the present moment, the likelihood of them being capable of meeting future obligations decreases. The related concept of "Caribbean time" is seemingly as prominent in the social sciences and cultural studies as is "Filipino time." In *Time Travel in the Latin American and Caribbean Imagination* (2012), Rudyard Alcocer recognizes a "unique significance" for time in the region. This unique significance is credited to a defiance of chronological linearity, whereby "Caribbean time seems to flout the linear 'clock time' associated with modernity" (Alcocer 2012, 68).

It is important to clarify that terms such as "Caribbean time," "Filipino time," and "African time" are not my inventions, nor even collective characterizations that I necessarily endorse. However, given the overwhelming presence of these terms in sociological and anthropological accounts of the socialized time structures of these respective regions, it is important that they are integrated within this discussion. I am wary that a term such as "Caribbean time" glosses over the differences in time structures of the various cultures and countries of such a region. Nevertheless, it is not incumbent upon the development of my argument to attempt to "unpack" the particularities of the component parts within such a homogenization. Rather, in recognizing the scholarly use of such characterizations (not to mention their reported utilization by the inhabitants of these regions), the key purpose for this stage of my inquiry is served in raising the notion of a globally pluralized construction of social time. If I were to attend to the differentiated politics between the component cultures or countries that comprise "Caribbean time," for example, this would in fact only further affirm this purpose. I am not though undertaking a socially scientific endeavor concerned with detailing different cultural/social forms of time. What is instead at stake is a philosophically critical excursion into the separation of such times from a natural tempo.

Not all theorists portray social time as a pluralism, but rather as a question of which cultures have constructed it versus those that have not. The concept of Caribbean time, and its connections to what is described as "African time," has, according to Joseph Adjaye's *Time in the Black Experience* (1994), been subject to either "denial or distortion by Western authors" (Adjaye 1994, 2). Adjaye's point is that in viewing non-Western time structures, in which temporal agreements differ from those of a Western author's context, judgments are made based on what non-Western frames perceivably lack in comparison to the West-as-default (1994, 2). In not having aspects of the mechanical minutiae of clocked representations of time, such cultures are often seen to be incomplete in terms of temporal sophistication and synchronization. For Adjaye, this demands that "Africans were depicted as being deficient in delineating relations of time" (3). The anthropology of Lucien Lévy-Bruhl is cited as a foundational example of this cross-cultural critique, which states that the "primitives' idea of time remains vague; and nearly all primitive languages are as deficient in methods of rendering relations of time" (Lévy-Bruhl, quoted in Adjaye 1994, 3). Lévy-Bruhl's assertion that certain African (and Melanesian) cultures have only vague conceptions of time is connected to his impression of their generally "primitive mentality" (Lévy-Bruhl 1923 [1922]). In raising Lévy-Bruhl, we can note the interest of his niece, the French philosopher of science Hélène Metzger, in reconfiguring this empirically developed portrayal of certain mentalities as primitive.

Lévy-Bruhl's thesis is actually strongly opposed by the chemist and philosopher Émile Meyerson, who argues against there being any real difference between a modern physicist's appreciation of the world, and the logic of supposedly primitive mentalities (Meyerson 1931, 31). Metzger disagrees with Meyerson's objection though, on the grounds that the latter's perspective glosses over the differences between human ways of thinking. To arrive at her own position, Metzger first recognizes a shared, homogeneous instinct, for all humans. Here she directs us to the notion of "expansive thought," which is described as the spontaneous, innate, unreflective mode of human thinking, which has not yet been formed into opinion or systemic logic (Metzger 1987 [1937], 35). By not excluding any humans from this "*a priori*" mode, Metzger destabilizes the reading that certain mentalities are more primitively instinctive than others.

Complementarily though, Metzger posits a plurality of humanly constructed forms of thought, which differentiate from the common, unreflective mode. There is, as a result, not merely one *a priori* direction for humans but multiple *a priori*, different, heterogenous, and incompatible orientations (Metzger 1987, 46). While these two phases are said to take the form of an "act" together, it must be said that within this characterization lurks traces of the distinction with which we have been working. I refer here to the separation of what is singularly naturally given, versus what is plurally humanly mediated. Despite this, Metzger's contribution is to unsettle a chronologically empiricist sense of primitivism. As Cristina Chimisso and Gad Freudenthal note, Metzger's sense of the *a priori*, while being "very close" to Lévy-Bruhl's impression of "mentality," differs in not establishing a developmental hierarchy between mentalities (Chimisso and Freudenthal 2003, 483). Similarly, Iris Van Der Tuin observes that assumptions regarding the "history of mentalities" should not be "mapped onto Metzger's own practice of thinking expansively" (Van Der Tuin 2015, 192).

Distinctions between supposedly primitive, versus industrialized, mentalities of time, are integral to the earlier kinds of discourses encountered that are concerned with globally differentiated regulations for lateness. Indeed, when Norbert Elias reviews the impressions of nonindustrialized time-structures, of the social scientist coming from an industrialized frame, he is not surprised that their conditioning makes it "hard to believe" that human beings exist without industrialized time-regulation (Elias 1992 [1987], 140). A culture's lack of a word for "late" is posited by Elias to be confusing for any social scientist who has grown up in an industrialized society in which it is ever-present (1992, 139). Capitalist, industrialist, social structures are characterized as standardizing mechanisms in how they divide and allocate time in order to maximize the efficiency of a cohesive workforce. The hallmark of industrial time is, accordingly, its partitioning into uniform slices. This serves two purposes. Firstly, it appears that punctuality, which is instilled as ethically virtuous,[1] becomes enforceable when time is divided into collectively agreed upon units. Synchronization is essential to a functioning workforce, with punctuality distinguishing the productive worker. Secondly, labor time, as a quantifiable component of production, can be used in commercial exchange. This is attributable to its standardization, whereby as Barbara Adam again observes, "the calculation of 'man-hours'. . . like the clock time units on which it is based, is an invariable, standardized measure that can be applied universally" (Adam 1990, 112). Standardized time orders the social body, providing greater control with more reliable outcomes.

In returning to Adjaye's argument, it is noted that non-Western social time structures with seemingly less strict lateness protocols, when viewed from a Western perspective, are consequently read as being less virtuously directed. What Adjaye demands, evoking Hacking's earlier definition of what defines a social constructionist's perspective, is an appreciation that no social time structure is the naturally inevitable version. Instead, for Adjaye, the pluralized construction of social time must be appreciated through an awareness of the cultural *relativity* of lateness, whereby "Africa and Black communities of the diaspora have to be studied and viewed as distinct products of their own environments rather than as aberrant forms of a Western norm" (Adjaye 1994, 8). Furthermore, regarding this plurality, and expanding on my earlier caution regarding the problem of homogenizing "African time," given the myriad manifestations of social time in African contexts, a singularized sense of African social time cannot be installed against a similarly reduced Western counterpart. This is a point on which Walter Johnson elaborates, demanding that rather than characterizing African time as a "single thing" (Johnson 2000, 490), inherent differences between African time structures must be acknowledged to reflect the "politically and historically embedded circuits through which they were transmitted" (2000, 491). While I have explained why it is not in the scope of my work to unpack these "circuits," these qualifications from Adjaye and Johnson are nonetheless worth raising in further recognizing a cultural plurality to time.

The point is hopefully clear: despite homogenizing perspectives, the plurality of social time becomes evident when recognizing its cultural relativity, which in particular we can examine through differing lateness regulations. According to one of the world's most distinguished scholars of African culture, K. K. Bunseki Fu-Kiau, for a society such as that of the Bantu-Kongo, the notion of lateness indeed cannot be accommodated at all. For the Bantu people "there is no such thing as being 'late' (unless they happen to have been educated outside Africa)" (Fu-Kiau 1994, 31). The Bantu are instead portrayed as having a present-oriented focus, where events in the here and now are prioritized over the abstraction of obligations beyond what is immediately occurring. What a differently socialized perspective might interpret in the Bantu as one's tardiness will contrarily be read locally as meaning that they have "been responding to other aspects" of the present "that were not foreseen at the time a 'fixed' point in time on the conventional time line was decided" (1994, 31). For Fu-Kiau, Western theorists must appreciate these shifting priorities not as quirks of a system lacking synchrony but as symbolic of a structure where what lies linearly beyond the immediate "is worthless" (31).

From such impressions again emerges the topic of the social constructionist separation of singularly natural and socially plural times. The plurality of socialized constructions is never more apparent, according to historian Mark Smith, than when considering African American integrations of, and retaliations against, "modern, clock-based time sensibility during slavery" (Smith 1997, 130). Here, in *Mastered by the Clock*, Smith describes how African Americans were able to "adjust to white time sensibilities, which stress punctuality and are future oriented" (1997, 130), while also rebelling against their masters by constructing their own rhythms within these sensibilities. This rebellion takes the form of "eschewing the authority of the clock and adopting presentist and naturally defined notions of time" (130). I flag that from this kind of analysis, we see the evocation of a "return" to a more originarily natural temporality that operates outside the jurisdiction of its culturally constructed forms.

Such theory makes the rudimentary point that there are different social structures of time, and that human, social phenomena do not illustrate naturally determined parameters. It is my assertion in response that advocations for a greater recognition of the culturally relative, globally multiple manifestations of social time do so by positing that African time, like Filipino time and Caribbean time, is distinguishable by its *present-centrism*. This characterization of present-centrism will be of considerable importance in considering critical reflections on the distinction between naturally singular time, and subsequent, representationally distant, social times. Influential social theorists such as Pierre Bourdieu perpetuate appraisals of present-centrism in arguing that the concept of the future is not intrinsic to an "African experience" of shared time. While social arrangements can be planned between citizens of such communities, for Bourdieu the consequent event is not understood to be occurring in the future. Bourdieu accordingly describes how "nothing is more foreign to the indigenous civilization of Algeria than the attempt to secure a hold over the future" (Bourdieu 1963, 55). Whereas for Bourdieu, the future in Western contexts occupies the conceptual realm of abstract, various possibilities, some of which might eventuate, for the Algerian community that he studies the future is that which will expectedly unfold based on what is already occurring or unfolding in the present. While this interpretation that African cultures lack the concept of a non-fatalistic future is not universally adopted, lax attitudes toward punctuality are typically, theoretically, attributed to a culture's perceived, greater present-centrism.[2]

Even in the Filipino context, tardiness is often equated with a population's collective lack of appreciation for anything beyond the material present. If we turn to the social anthropology of Melba Maggay, Filipino time is described as less concerned with the abstract coordination of what is not yet occurring and more centered on one's "present orientedness" in which "events are confronted only when they present themselves" (Maggay 1998, 370). Indeed, the fact that for "Filipinos, time is not sequential" (1998, 371) is in this manner celebrated by anthropological accounts of time seeking to culturally relativize various, globally situated, time structures. A time structure such as that found in the Philippines, in which the focus is not always on a "mechanical succession of minutes and seconds" (371) is used to illustrate how highly functioning societies can operate without what other cultures might take for granted concerning social synchronization. Consequently, even if, as Jan Selmer and Corinna de Leon observe, "among Filipinos for whom time is elastic and deadlines are absent, a strict sense of punctuality, regularity and timekeeping are difficult to instill and enforce" (Selmer and de Leon 2002, 12), a role is proposed for sociology and anthropology to broadcast an appreciation of how such societies still function effectively. This speaks to the responsibility with which such disciplines are invested in guiding us beyond what we take for granted about the socially constructed version of time in which we are embedded. In appreciating the plural differentiation of social time structures by arguing that none of them were "naturally inevitable," Harry Bash argues that what requires interrogation is the belief in a default or hierarchized social time structure, otherwise a certain "tempocentrism follows as 'naturally' from the construction of social time as ethnocentrism attends to the social construction of culture" (Bash 2000, 196).

Such tempocentric assumptions and hierarchies regarding social time are also acknowledged by Anthony Aveni. Through processes of globalization,

Aveni notes that many cultures "are united by time zones, an international dateline, and a universal second" (Aveni 1989, 338). Cultures that remain outside this frame are seen to be incapable of participating in it because their temporalities "have been organized in less complicated ways" (1989, 12). "Complexity" here becomes reducible to a model that transcends the immediate present. Aveni thus reiterates the characterizations just examined, in declaring that for African time structures that operate outside the global standard, "linear connections, causality, and sequentiality are no concerns of theirs. They place no value on the future as we do when we use the word progress" (332). While a commentary such as this seemingly indicates a tendency toward a theory of cultural hierarchy rather than of cultural relativity, Aveni is quick to acknowledge that linear time is not exclusively a "property of advanced civilized societies" (333). Contrarily, the notion of "advancement" is always "value-laden" (333), meaning that structures of linear time are entirely contextual and largely incomparable.

Despite this appreciation of the culturally contingent construction of time, the network of international time zones that Aveni mentions installs a hegemonic Western conception of time as the global default or ambition. Its influence means that the impression time-theory conveys, in returning to Nowotny's commentary, is of the evolution "of one dominant (the Western) concept of time, out of many highly different 'local times'" (Nowotny 1992, 426).[3] Quite interestingly for my intention to interrogate the seemingly polarizing conditions between natural time and social time installed by aspects of social constructionist thought, in considering to what this perspective is attributable, Nowotny suggests that *some social times are considered to be more natural* than others. This occurs via the theoretical juxtaposition of (a) clock-based, social time structures from (b) the various forms of social time without such symbols. From this distinction, the latter time structures are judged to be "more 'natural,'" and to result "from adaptations to seasons or other kinds of natural (biological, environmental) rhythm" (1992, 428). This interpretation, that different social time structures are potentially being qualified according to a perceived proximity to an original, natural state, marks a pivotal feature of my interest in this inquiry. While I will outline key aspects of this understanding now accordingly, this thesis will be an ongoing point of discussion as philosophical interventions are later integrated that destabilize distinctions of natural and representational temporalities.

THE PROXIMITY OF SOCIAL TIME STRUCTURES TO NATURAL TIME

That certain social time structures could be considered to be more or less natural evokes the social constructionist model with which we are working in which a singular, natural phenomenon is splintered into myriad social representations, each trying to (re-)access the real/natural phenomenon. For Nowotny, as the ramifications of pluritemporalism emerged in the physical sciences via Einstein's contestation to Newton's absolute time, so a range of disciplines in the humanities and the social sciences felt they were "free to posit the existence of a plurality of time, including a plurality of social times" (Nowotny 1992, 428). Prominent theoretical physicist Sean Carroll equally emphasizes the interdisciplinary legacy of this scientific development. In the view of Newton, there was an absolutely correct way to slice up the universe "into splices of space at a particular moment of time" (Carroll 2010, 11). Conversely with Einstein's theory of relativity, what became apparent was that two clocks will not always experience the same duration, whereby "the real world . . . doesn't let us construct an absolute universal time coordinate" (Carroll 2010, 11). The point of this observation for the topic at hand is that as the social constructionist position that I have identified posits a multiplicity of contingently representational "clocks" of time, the question is whether some of these representations are believed to be closer to what is still held to be naturally real about time.

Nowotny would seemingly respond in the affirmative. The conception of time as an externally flowing phenomenon, as per Newton, is, according to Nowotny, assumed to be "an external reference frame, against which 'social time' could be posited" (Nowotny 1992, 426). In returning to critiques of Caribbean time, African time, and Filipino time, the distinction of all three from clock-time structures in which lateness protocols are stricter has been, in part, portrayed in terms of a naturally material, present-centrism. For Katrin de Guia, this is explainable in terms of Filipino time because of it being a "cosmic time," *not* a "clock time" (Guia 2000, 187). Filipino time is indicated as having a greater connection with natural events than with abstract representations of natural temporalities. This is further reinforced in Guia's claim that in the Philippines, what is experienced is an "organic time" that is "cyclical, oscillating . . . alive," in contradistinction to the "repetitive staccato" of Western machine time (2000, 187). I interpret that organic in this definition correlates with the aforementioned notion of the event as that which is actually, presently material(izing).

The perceived resilience of such a time structure to remain in coherence with natural events and natural rhythms, despite the global encroachment of mechanical, clock-based time structures, is equally portrayed in conceptions of African time. We have already reviewed Mark Smith's characterization of the way African American slaves maintained a sense of ecological, rather than of modern industrial, temporal rhythms. While Smith reports that African social time structures have a greater degree of ecological alignment than those of the West (Smith 1997, 133), theorists such as Douwe Tiemersma go to the even greater extent of defining modern African time concepts in complete juxtaposition to Western ones. Tiemersma duly describes an "organic time . . . in opposition to Western time defined by clock and calendar" (Tiemersma 1998, 269). A pattern emerges regarding conceptions of these temporalities accordingly, in which social time structures functioning outside clocked and calendared time are depicted as lacking strict lateness protocols and as being more present, or close, to naturally organic temporalities.

Beyond accounts concerned with particular cultures, in returning to the work of Anthony Aveni we encounter an impression of how a conceptual divide between natural and human times pervades *all* arenas within the humanities, including poetry and literature. Through the imagery provided by Hesiod, the description of time as "the ordered cycle of sensible natural events to which human beings were meant to relate the events in everyday life" (Aveni 1989, 51) becomes apparent. Indeed for Hesiod, the essence of time is identifiable in a continuing dialogue between nature and culture whereby the latter is supposedly a perpetually shifting approximation of its singular, unchanging origin. A sense of the subsequence of social time to natural time is here apparent. Natural time represents a resource that we can symbolically manage for our own purposes, where for Aveni we "manipulate nature's direct input into the timekeeping process for our own benefit" (1989, 337). Celestial time for Aveni (or what David Hoy describes as "the time of the universe" [Hoy 2009, xii]), exists first, then it is (mis)represented by a multiplicity of cultures. After this, social time and natural time are observed to be only conditionally integrated together in daily rhythms. This "human intervention," Aveni notes, merely comprises a subsequent "insertion of society's time into celestial time" (1989, 337).

Social theorist John Urry confirms these polarizing characterizations of time when he states that most explanations "have presumed that time is in some sense social, and hence separate from, and opposed to, the time of nature" (Urry 2000, 417). I believe a definitive example of this distinction is sourced in Pitrim Sorokin and Robert Merton's early sociological perspective on time: "Social Time: A Methodological and Functional Analysis" (1937). In this work, the time found in human societies is distinguished from the unerring cycles of seasonal, natural time, with the latter being described as "uniform, homogenous; it is purely quantitative, shorn of qualitative variations" (Sorokin and Merton 1937, 621). This influential investigation informs, to some extent, the aforementioned social constructionist tendencies within sociologies and anthropologies of time, defining a "system of time which varies with the social structure" (1937, 621), in contradistinction to the invariant, objectively constant, time of nature.

The argument I have been developing through this analysis of literature concerned with the relation between natural time and social times is that natural time is presented as a blank, unified, present, and homogenous slate against which the constructed complexities of different social time(s) are compared. The unerring regularity of natural time is defined as that which is subsequently *used* socially, underpinning the characterizations of time as a resource. Sorokin and Merton agree, for "where natural phenomena are used to fix the limits of time periods, the choice of them is dependent upon the interest and utility which they have for the group" (Sorokin and Merton 1937, 621). This utility or interest, according to the social constructionist position being targeted in this inquiry, is interpreted to be secondarily contextual and not an indication of time's "natural order." Barbara Adam's research into the construction of conceptual distinctions between social time and natural time complements this key claim that typically "social time seems defined against 'an other' which appears to be a convenient backcloth" (Adam 1990, 150). This divides social times from the natural cycles on which they are based, such as night and day, birth and death, and harvesting seasons.

Despite the conceptual separation of natural time and social times, what must also be considered is how social time seemingly *accommodates* natural rhythms and temporality. A most significant characterization of time in this sense is found in Edward Evans-Pritchard's study of the African *Nuer* tribe. Evans-Pritchard documents the dependence of the Nuer's social temporality on the natural cycles of the "cattle clock" and other seasonally based activities. The argument is that in such an environment, collective activity *is* time, whereby Nuer time never becomes abstracted as a commodified resource. Their language apparently illustrates this, for the "Nuer have no expression equivalent to 'time' in our language," preventing the commercial, exchange-driven, quantification of time, as something "which passes, can be wasted, can be saved" (Evans-Pritchard 1940, 103). Points of/in Nuer time are "not controlled by an abstract system" (1940, 103), but are identified as deriving from seasonal, social activity, and the associated ecological cycles. Given this correlation between social activity and temporality, Evans-Pritchard concludes that during periods devoid of such activity, the Nuer do not refer to time at all.

By defining time as "the significance which natural changes have for human activities" (Evans-Pritchard 1940, 104), Evans-Pritchard differentiates, then synthesizes, natural time *and* Nuer social time. This differentiation and subsequent combination, a two-step process as he presents it, is designed to accommodate both the Nuer's "relations to environment, which we call oecological time, and their relations to one another in the social structure, which we call structural time" (1940, 94). My reading, however, is that in spite of the dual constitution of Nuer time, what is still posited is natural time as anterior, it exists before it is restructured via seasonally directed, social activity. Crucially for our current considerations, this structural activity is conceived to be less distant from the natural origins of time than the social time structures found in industrialized or commercialized contexts. Evans-Pritchard's argument that the Nuer lack time-vernacular/words evidences this, assuming Western clocked and calendared time notions as detached constructions that are not present in Nuer time. Note his astonishment that the Nuer "do not use the names of the months to indicate the time of an event" (100). Nuer time, for Evans-Pritchard, a primal unification of "oecological cycles" and structural relations, is according to this interpretation given a certain, socialized status through a comparison of it against the greater abstraction and symbolization of the Western clock and calendar.

Such clocks and calendars, in harnessing and mediating natural temporalities, are viewed to operate further from ecological rhythms accordingly. The social time structure that we might absent-mindedly take for granted as naturally being time, is conversely installed as a fabrication. Reinviting Aveni to the discussion, the impression is imparted that "time—we have socialized it, circularized it and linearized it, artificialized and corrupted it" (Aveni 1989, 85). Indeed, this supposition of the artificialization of what is naturally real about time is further developed in Aveni's appreciation of the contingently different ways that time structures emerge globally. What his research of various, historical, social time structures, recognizes "is that time captivates not as a pure fact of nature but instead as a dimension of life that ultimately can be submitted to cultural control" (1989, 336). As I have earlier asserted, when a time structure is believed to be more culturally abstracted from natural, celestial rhythms, such as in the mechanical, industrial West, then its "control" of lateness is also deemed to be stricter than that of present-centric time structures. An extreme ramification of this is not just that less-strict lateness regulations will be associated with present-centric time structures, but that the associated cultures will be viewed as less developed. The lateness protocols embedded in Filipino time, as Joseph McCallus notes for example, are judged from a globalized frame in which the "Western attitude of promptness is deemed superior and the Filipinos are culturally inferior because they view time rather differently" (McCallus 1994, 51–52).

This brings into focus the emancipatory efficacy of time-theory. Social constructionist assertions, which ask us to reevaluate our "tempocentric" perspectives about which cultural structures are deemed to properly structure a "socialized time," seemingly perpetuate, rather than disrupt, structural oppositions between natural time and social times. In relativizing social times through the constructionist argument, a symptom emerges where time structures in which lateness is less regulated and a present-centrism dominates, are often marginalized from a developmental temporality that is associated with humanity's "break" from nature. The question in this book concerns how the philosophical methods that follow in the coming chapters intervene into the aspects of methods that perpetuate this constructionist condition.

This theme, of the sequential separation of originary natural rhythms, from subsequent, social, contingent rhythms, can be further opened through the context of human, corporeal rhythms. The reason this discussion turns to such a context is that in considering the intersection of socialized lives with natural rhythms, the body manifests as one's personal site of what is perceived to be "natural time." Just as the movements of the celestial bodies are viewed by scientific realists to indicate a temporality that has a truth beyond socially contingent constructions of it, likewise our corporeal temporality, and eventual trajectory toward corporeal death, might be perceived to exhibit a physical fact that is independent of any social impression of it.[4]

Consistent with the argument observed to this point that assumes a prior, natural phenomenon that is then socially constructed in different forms, through examining the corporeal politics of body modification practices, we can attend to how the natural body is seen to be socially (re-)constructed. What is central for our concerns in such a discussion is the distinction that might continue to be exhibited between natural and socialized temporalities. A pertinent question to this end is whether these socially prescribed practices arrive late on the scene of an already originated, natural entity. Readers might feel a personal investment in such debates, as the study of bodily temporalities induces reflections about the timing of, and one's relation to, their own bodily changes.

Examining assumptions of the temporal relations between natural bodies, and the socialization of those bodies, also opens the politics shared between human bodies and theory about bodies. As material human bodies are theoretically and socially/culturally represented, which in turn repositions material bodies via different impressions of those bodies, what must be considered is the relationship between the material present and the representational subsequent. Is there a simple sequence of bodies first originating and existing as natural entities, then socially constructed versions of these bodies (both physically modified, and theoretically constructed), manifesting subsequently or secondarily? Are human bodies physical/material, but not the bodies of theory about bodies that are nevertheless so implicated in physical/material bodily construction? In order to explore this interchangeable dimension of physical human bodies, and social constructions of those bodies, it is necessary to firstly become familiar with the body modifying context.

MODIFYING THE NATURAL BODY

Definitions of body modification practices often portray the individuals who undertake them as agents who intend to alter their corporealities. Sociologist Mike Featherstone observes that within the category of body modification are "practices which include piercing, tattooing, branding, cutting, binding and inserting implants to alter the appearance and form of the body," as well as "bodybuilding, anorexia and fasting" (Featherstone 2000, 1). Bodies are assumed to change "naturally" when something like the gradual wrinkling of the skin represents the prolonged, inevitable, temporality of aging. There is a different appreciation of the change from pre-tattooed to tattooed skin, however, where the limits of corporeal time become dramatically, suddenly conspicuous.

The voluntary modification of the body has been identified in human societies spanning thousands of years. The anthropological research of Gloria Brame, William Brame, and Jon Jacobs discusses at length how "historically, travellers' tales and the works of anthropologists have shown that body modification is virtually universal" (Brame, Brame, and Jacobs 1993, 298). Tattooing, one of the most ancient and widely undertaken forms of body modification, is said to provide an illustration of this relative universality. Archaeological findings indicate that tattooing was commonly practiced in all eras of the Stone Age for which we have evidence. Clinton Sanders and Angus Vail note that "carved figures from European sites dated 6,000 years B.C., and Egyptian figurines created some 2,000 years later, show facial and body markings thought to represent tattoos" (Sanders and Vail 2008, 9).

Tattooing in ancient Egypt is believed to have been restricted to women, particularly dancers and priestesses.[5] Conversely, there is evidence in other cultures of tattooing being used to indelibly mark, identify, and marginalize social deviants. In sixth-century Japan, for example, criminals and social outcasts were tattooed on the face or arms as forms of public identification and punishment (Richie and Buruma 1980, 12–13). This contrasts with the high status attributed to the moko tattoos on the lip and chin areas of Maori women, and to the extensive geometric facial and body tattoos of Maori men. These designs are so indicative of the individual's social role that "following contact with Europeans, they [the tattoos] were often used by members of the nobility as signatures on legal documents" (Sanders and Vail 2008, 10). Body modification practices thus take on different significations, contingent on the social space, context, or era in which they are undertaken.

The modes of body modification that are accommodated and accepted in a modern, Western context often comprise those by which a more youthful appearance is anticipated. In observing that regularly in such social spaces, notions of youthfulness and beauty connect, for Gloria Brame these practices become customary because "Euro-American culture has esteemed modifications that reverse or stall the effects of aging" (Brame et al. 1993, 301). Here we see how cosmetic surgeries that remove wrinkles become normative and acceptable. Victoria Pitts's[6] *In the Flesh* (2003), a text to which we will again refer, further illuminates this discussion concerning the cultural conventions of body norms in noting that such surgeries are "not only acceptable, but almost expected of people of a certain gender and class status" (Pitts

2003, 35). Conversely, modifications such as scarification and branding are commonly criticized. Scarification involves professionally applied knife cuts, whereas branding burns patterns into the skin. These practices are often seen as *violently* self-mutilating rather than as self-beautifying, to the extent of being, as Pitts's research reveals, "linked to anorexia, bulimia, and what has been called 'delicate self-harm syndrome'" (2003, 25). Both are often aligned with body piercing and tattooing, as affronts to the aforementioned Western beauty norms (15).

A different appreciation of scarification pervades certain African contexts, however, where it has been traditionally, symbolically employed. Women of the Sudanese Nuba tribe receive cuts that mark their physiological maturation, in that, as anthropologist Robert Brain notes, cuts are made at puberty, at the onset of menstruation, and after the woman's first child (Brain 1979, 70–73). Once scarification has commenced, one's membership to a particular tribe is symbolized, as is their passage into adulthood. One must be cut to belong. This contrasts with the institutionalized critiques of the same practice in a contemporary, Western context, which portray the body modifier as socially distanced. Sanders and Vail's research notes this typical characterization of scarification in North America, the media positioning the practice as "a body alteration eminently suited for symbolizing disaffection from mainstream values" (Sanders and Vail 2008, 8).

This collective criticism of non-normative body modifications can perhaps be traced to legacies identifiable in the earliest recorded fragments of Western philosophy concerned with what constitutes the borders, or limits, of someone's or something's essential or natural identity. The pre-Socratic philosopher Heraclitus famously wonders in relation to change how a thing can remain the "same thing" while becoming different.[7] According to Bertrand Russell, such curiosity is not simply an inquiry into the nature of change, but rather reflects the human desire for stability (Russell 1961, 63). Bruno Latour similarly recognizes stability's prioritization. In *We Have Never Been Modern* (1993), Latour proffers stability as one of two variables that connect Nature and Society. In this sense, the more stable something is about the world, the more identifiable it is to an inquirer. Even when identifying change, the human is understood to be searching for something that transcends, and inherently conditions, such change. This theme will resurface throughout this book in terms of an ongoing analysis of conceptual differentiations of naturally necessary, and socially malleable, phenomena. Internal to this reading is that if a social phenomenon is a contingent construction, then it is only by tracing it to a natural foundation that a true meaning can be discovered. For the context of body modification, this positions the physical body as the real ground from which variously less-real corporeal timings emerge, including modifications as markers of avoidable change. As we will see, these socially informed and framed markers are often criticized for being distanced from the supposed, inherent, truth of the body, defined against, as Pitts states, "implicit assumptions about the body as naturally pristine and unmarked" (Pitts 2003, 75). It is with such characterizations that suppositions emerge regarding socially informed practices arriving late upon an already established, natural phenomenon. Bodies, as exemplars of the necessary cycles of a natural temporality, in changing and becoming "modified bodies," are abstracted from a natural state via the constructed marks of social contingency.

In tracing a history of philosophical thought on relations between time and change, the first recorded, surviving Western philosophical conception of time is identified in a fragment attributed to another pre-Socratic philosopher, Anaximander of Miletus. Anaximander posits the source of all things in the world as "some other *apeiron* nature, from which come into being all the heavens and the worlds in them" (Anaximander 1983 [circa 550 B.C.E.], 118; author's original emphasis). This apeiron nature is an unlimited, unchanging, source of all worldly, changing, or limited things. By regulating the limited nature of worldly phenomena, time functions as justice. This just "assessment of Time" (1983, 118) means that the authority of an unchanging, anterior, divine realm, is distinct from the perishable realm of changing things, such as the human body. The source of the time of bodies, and of the body itself, is positioned outside the realm of bodies. I suspect that the pathologization of certain body modifications, which features the "idealization of the natural body in self-mutilation discourse" (Pitts 2003, 33), presents a potential correlation between the interpretation of an unchanging divine nature and the divinely created, pre-modified-by-the-world, human body. As will be explored in later discussion, with the introduction of body modification practices, human corporeality is seen to morph into an "improper," socially constructed, version of its true or natural incarnation. What we need to be aware of is how, just as social time is theoretically constructed to be a contingent symbolization of an externally anterior, naturally worldly, tempo, the socially modified body is also portrayed as a contextual reconstruction of its natural limits.

The social constructionist impression of a "real source" that is inaccessibly separate from the subsequently contingent, worldly representations of it, could in this manner be partly informed by perspectives found in classical mechanics. The Italian physicist Galileo Galilei represents time geometrically according to the regularity of the displacements of distance (space) of bodies in motion. When measuring a particle with uniform motion, it is stated that "the distances traversed by the moving particle during any equal intervals of time, are themselves equal" (Galilei 1914 [1632], 154). Time, in being represented geometrically, is here defined by a uniform regularity whose conditions are outside the motions themselves. This exteriority from the interference of a changing, material realm characterizes time as an objectively independent variable in the measure of motion. Rather than focusing on the notion of a time source that is outside material change, if we highlight the associated belief in a "real time" that is objectively separate from the measures and representations of it, we see how this perspective might inform the social constructionist time condition.

The influential English mathematician Isaac Barrow also attributes the uniformity of time to the exteriority of its divine source from the materially spatial realm. Barrow explores whether time existed before the creation of the world (Barrow 2009 [1683], 160). Time, if independent from worldly interference, does not require worldly change or motion for it to exist. Given that time is conceived by Barrow as flowing *evenly* outside worldly interference (otherwise time[s] would not be comparable), not just any motion can distinguish time. Rather, the measurement of time requires *uniform* motion, which "proceeds always in an even tenor" (2009, 160). Barrow recognizes the uniformity of the celestial motions accordingly as distinguishing the measuring of hours, days, weeks, months, and years. From each of these uniform "parts of an equable motion" (160), time is represented as an externally governed, straight line. I interject here to note that the ramification for body modification discourses of linear, unidirectional, forward-moving, objective time, whose source transcends the embodied, human realm, is that the pre-modified body is permanently installed in a separate, previous past at a point to which the body can never return. Socially contingent modifications make the natural body irretrievable, or as Margo DeMello observes, are "permanent and alter the body forever" (DeMello 2007, xvii).

A variation of Barrow's model was also to be developed by one of his students, physicist Isaac Newton. Newton maintains, with Barrow, that the source of time is objectively outside human interference but within the realm of God. From this objective separation, a "true, absolute time" manifests that is uniform and regular, because it "flows equably without relation to anything external" to itself (Newton, Hall, and Hall 1978, 77). Given that such time is absolute, Newton posits that there is an objective succession of phenomenal transition, whereby "all things are placed in time as to order of succession" (Newton 1952 [1673], 79). This continuous temporality objectively conditions the material/spatial/physical world, whereby we could say that it must sequentially order and permanently position something like the pre-modified body prior to its modified counterpart. The social constructionist position would accordingly be to appraise this modified counterpart, the socially produced contingency, in terms of its lack of inevitability. It arrived secondarily and will be superseded in turn.

True, absolute time also *appears* to be able to explain how humans socially synchronize. Individuals synchronize with each other by adhering to the same representation of an original time. That original time, according to a Newtonian perspective, maintains its uniformity and regularity for all individuals by transcending the interference or whims of any particular individual. We have acknowledged that Albert Einstein's theory of relativity (Einstein 2006 [1920]) complicates the possibility of a singularly, absolutely objective, time. Nevertheless, within localized settings, I posit that a powerful assumption emerges from Newton's thesis that is relevant to the impetus of this study. Such relevance is attributable to how this position captures a perspective on time that humans seemingly take for granted; that there is an objectively separate rhythm to the universe that rolls on relentlessly, regardless of individual desires. This informs the belief that one cannot affect the rate of time to suit their own needs and that time will flow at the same rate that it always does, no matter what the individual requires of it. If someone is bored at work and tries to make time go quicker, what is actually attempted is to make their experience of time different, qualified by the commonplace concession that time itself will actually still occur at the same rhythm. The social constructionist time condition holds that representations of time merely use this unerring rhythm to condition social functions such as collective synchrony.

The ramification of these positions on time for body discourses is the assumption that a corporeality that moves and changes state according to a "constant and uniform" temporality (Newton 1952, 376) will age consistently. Sociologist Wilbert Moore evokes this assumption in stating that "age, being basically a function of time, should move at a steady rate" (Moore 1963, 60). Empirical support for this claim is found in the research of anatomist Harold Brody, which reports that neurological studies on brain degradation illustrate that "brain weight decreases continuously at a uniform rate, the maximum decrease between 25 and 96 years being about 11% of the mean weight of a series of 2080 brains" (Brody 1955, 512). Such medical studies form part of greater social discourses that assume that the normal body ages, or changes, regularly over time. Even abnormal bodies, such as those afflicted by the rare genetic condition "progeria"[8] (in which the physicality of "old-age" manifests at an early age), are seen to change at a constant, albeit comparatively accelerated, rate (Brown 1992, 1222S). What is therefore being illustrated is that the body's dependable aging, in normal and extreme forms, is presumed to obey the unstoppable, inaccessible, force of objectively separate, uniform time.

Given how uniform time has been associated with the regular movements of celestial bodies, a conceptual association emerges between the spatial movements/transitions of these planetary bodies, and those of the human body/corporeality. Or as Gail Weiss notes, given that the temporality of clocks and calendars is based on the materially spatial (corporeal) movement of the Earth around the sun, there seems to be an aspect of clock time that is "not merely an external, analytical device that helps us negotiate our everyday affairs, but is based on corporeal movement, movement that is inscribed in our own bodies" (Weiss 1999, 112). One way in which this inscription manifests is seemingly that the apparent uniformity of planetary rhythms is correlated to the human body, whereby all such bodies are presumed to spatially or physically transition over time at a constant rate.

The intended goal of this brief scientific commentary is a wider appreciation of what could inform modern conceptions that install a real phenomenon, anterior to its socially constructed (modified) form. The body modifier, in conflicting with a regular temporality that supposedly emanates from either a divine realm or a celestial realm, and that conducts all such bodies, is duly acting in sin. This is seen in the anti-tattooing sentiments displayed in Leviticus, declaring "you shall not make any cuttings in your flesh . . . or tattoo any marks upon you: I am the LORD" (Leviticus 19, 28). A similar valuation of the body features in the New Testament's *First Epistle to the Corinthians*, which asks, "[D]o you not know that your body is a temple of the Holy Spirit, who is in you, whom you have received from God? . . . Therefore honour God with your body" (1 Corinthians 6, 19–20). What I posit is a connection between theistic conceptions of the body, and social discourses that criticize the secondariness of certain body modification practices. Both positions conceive of body modifications as contingent after-effects of subjectively *constructed* anomalies and define such voluntary intrusions upon human corporality in terms of their *destructively untimely* moralities. The earlier encountered, conceptual opposition, dividing lateness from virtuousness, possibly reemerges.

The discussion to this point has tracked arguments in which a natural, celestial phenomenon is conceptually installed anterior to its social derivations. By investigating the relative, culturally plural structures of social lateness, the social constructionist condition regarding time has been evidenced as relying upon, and indeed installing, this natural|social divide. Naturally real time precedes representationally social times. Now, when the theme of lateness has been opened through the context of body modification practices, what we see again are assumptions of a natural phenomenon that is *secondarily* socially fabricated in myriad, contingent ways. The natural body is conceptually installed as that which precedes social bodies, whereby according to the characterizations just appraised, the socially modified or constructed body marks a genuine end, or even some kind of death, of the natural purity of the body. As the socially modified body emerges, so the natural body is considered to have passed.

What manifests is a double sense of lateness to which the context of body modification practices speaks. In one regard, we have seen perceptions of a literal secondariness, or a subsequence, to the socially derived, modifying practices that arrive to, intrude upon, and interfere with an already originated and established, natural phenomenon. That is, the socially modified or reconstructed body arrives late on the scene of the natural body. In a complementary regard though, as the socially modified or reconstructed body is born, in an interchange that is perceived to simultaneously signal the end of the body's natural purity, a certain mortality for the natural body is heralded. It is as a result of this *mortality* of what is perceived to be a natural origin, that it is the natural body which resonates with connotations of being *late*. This second relevance to the theme of lateness in terms of the socially constructed body emerges via significations of death.

How can such a claim be made, that despite the socially reconstructed body arriving after, or secondarily to, an already occurring natural body, that it is counterintuitively the natural body that can be described as late? Or in other words, how can a body, which is deemed to be synchronous with a natural state that *precedes* untimely social divergences, be the late body? In thinking through this twist, we must revisit the social constructionist position with which we are primarily engaged, in which the natural conditions of socially contingent constructions or productions are considered to be inaccessible to those social constructions and their agents. Social phenomena according to this position do not capture the truth of a natural reality. Rather, they are viewed as approximations of an inaccessible, out-there in the world, reality. This reality is inaccessible because of the inescapability of our contingently constructed perspectives of it. All that we ever know, all that is ever alive to us, is what is socially constructed.

This means that as the socially modified or reconstructed body manifests, the mortality of the natural body is not even a "development" that we witness. For the social constructionist, we do not have a prior access to the natural phenomenon, before we, and it, become shaped and shifted socially. Rather, all that ever occurs to us corporeally is a socialized body. We only experience the body, conceive of the body, modify the body, and construct the body, through socially fabricated lenses and senses. The modified body could be said to be less of a *re*construction, and more simply, originarily, a construction. The natural body is never alive to us, and yet we mourn and attempt to (impossibly) revive/revisit the late, natural body. Through practices including paleo dieting, body cleanses, and seaweed treatments, humans go to considerable lengths to return to, or catch a corporeal glimpse of, an original, natural, pure state. However, the idealized return is always conditioned by socially constructed compositions and perspectives, that for the constructionist thesis make any such "complete return" impossible. Cohering with the definition that equates late with dead, the natural body is late because it is not alive nor arriving.[9]

Aspirations, and considerations, of a return to pure experience, bring with them tones of phenomenological reductionism and the "bracketing out" of one's judgmental fabric. This is not a coincidence, given that in chapter 5 the existential theme of mortal lateness will be framed phenomenologically when examining the fear that we are running out of time, or are already too late, to save our species from catastrophic, environmental change. Of more immediate concern, in the preceding paragraphs we have been occupied with the literally fleshy construction, the physical modification, of the socialized body. Implicated though within all such considerations are the theoretical constructions by which we come to develop a sense of the body. The body is socially constructed, at a perceived distance from its original state, through the *physical* markers of tattoos, diets, and plastic surgery. However, it is also constructed *theoretically* through ideas about the body, which come to inform, and be informed by, the physicality of the body. Within the social constructionist time condition that installs a sequence between a physical reality, and the constructed representation of that reality, is the interpretation that first there exists the physical body, and then there are subsequent, socially constructed, theoretically representational, forms of it. Nevertheless, could the earlier play on the double meaning of the word lateness, which posits that the natural body is the relatively late entity despite the socialized body arriving secondarily, be an indication of what is not straightforwardly sequential between nature and culture for social constructionist temporalities?

Consistent with this last thought, I want to consider how, despite social time structures appearing to be culturally contingent subsequences of an objectively separate, natural phenomenon, there is actually something objective about a social time structure itself. This is apparent in the everyday reality of social synchronization and function. We can hence consider what a foundational theory of socially objective time could entail for the constructionist position of a source of time that is objectively separate from subjects. Parallels between the temporalities and constructions of social bodies and human bodies will continue to guide us.

THE OBJECTIVE (SOCIAL) FACT OF TIME: DURKHEIM'S STRUCTURALISM

The structuralism of Émile Durkheim can be considered to be a precursor to contemporary social constructionist positions on socialized temporalities. A pivotal era of Durkheim's structuralist outlook in this regard examines suicide rates in the late nineteenth century. What will be of interest to us is whether such suicide rates, which Durkheim identifies as having reliably predictable and even uniform rhythms, are exemplary of the regulation of human tempos by an objectively inaccessible time-source.

Durkheim's attention is in fact on what he describes as being "objectively coerced" about one's socialized temporality. Even though Durkheim therefore diverts from our considerations about an objectively external *natural* time, his focus still speaks to our current theme. This is because, despite the earlier recognitions of a relativity of time and lateness protocols between different cultural and corporeal settings, we can note that embedded within any particular social time structure are impersonally preexisting, rather than subjectively personalized, roles and expectations about time and lateness. To avoid being late does not simply mark one's individual timeliness but is where one expresses a temporalizing force that predates one's existence and that might even seem timeless.

This regulation of individual tempos materializes in Durkheim's intrigue concerning how each social period produces a regular suicide rate. Attending to the *social* rhythm of such behavior contradicted the typical assumption that suicide was simply individually authored, Jack Douglas noting Durkheim's differentiation from views that suicide is "an intensely individual act" (Douglas 1967, 16). Just as there is a tendency to define suicide in terms of individualism, similarly a discourse of autonomous causation pervades our ongoing context of body modification practices. As DeMello notes, the usual interpretation is that "individuals choose to take their own bodies into their own hands, and modify themselves via piercing, tattooing, surgical and pseudo-surgical practices" (DeMello 2007, 32). The notion of the "body project"[10] for the body modifier is paramount, whereby particularly with tattooing and piercing, Pitts's research reveals a perceived "self-control over one's body through self-inscription" (Pitts 2003, 10). This is further evidenced at the central hub of the body modification community, the website *Body Modification Ezine* (*BME*), where Pitts recognizes "the highly individualistic discourse of *BME*" (2003, 169). The interpretation of suicide as a sovereign process, against which Durkheim will argue, is here replicated in discourses of autonomy associated with body modification practices. Indeed, if we recall that body modification practices are commonly identified as one's self-harming, and as manifestations of self-destructive subjectivities, the contentious link between body modification and suicide is expanded.

Equally, an interpretation of social lateness is that an individual's transgression of collective temporal agreements indicates that they operate at their own temporality, *outside* or defiant of social influence. Perhaps most importantly for us is that this reading, of the potential separation of an individual's time from socially structured time, further develops the idea of sequential time conditions. Previously we have encountered the presumption of sequential temporalities regarding the relationship between natural time and social time. First, there is a singular natural time, then there are socialized constructions of it. In this current consideration, however, the focus is on whether a social time structure's regulation, which is portrayed as a singularly true temporality for the individuals within it, precedes the subjective timings of those individuals. Such a theme is analyzed extensively in the following chapter on subjective time.

Regarding our current concern with what might be objectively regulated about socialized temporalities, in challenging presumptions of subjective causation in suicide, Durkheim adopts what he describes as a scientific approach. By interrogating the concept of an isolated individual subject, he instead discovers the statistical regularity of suicide rates, suggesting a collectively shared, rather than a personally motivated, basis for any individual manifestation of suicidal behavior. This is important in terms of the capacity of the social scientist to study suicide, because as he expresses, "a scientific investigation can thus be achieved only if it deals with comparable facts" (Durkheim 1952 [1897], xxxix). The individual is only relevant as a particular aspect of a comparable, greater whole, in this outlook, reflecting Durkheim's interest in the broader, social production of the subject/self.

When considering such production beyond suicide, Durkheim believes subjects are always informed by a past that is alive in the present. One learns to become social, and while the source of this social consciousness is objectively outside the individual, such education is unavoidable for all individuals (Durkheim 1938 [1895], 6). By realizing how populations behave predictably and regularly, the causes of individuals' general behaviors are traced to structurally durable, rather than to individually idiosyncratic, conditions. These conditions are identified in *The Rules of Sociological Method* as roles that manifest via individuals as "ways of acting, thinking and feeling that present the noteworthy property of existing outside the individual consciousness" (Durkheim 1938, 2).

The reason for exploring this structuralist view of social regularity is to integrate its influential sociological impression that the rhythms of human subjects are regulated by a temporality, the objectivity of which requires a source that is inaccessibly external to those subjects. Unlike for the social constructionist time condition being investigated though, this externally real source of time for Durkheim occurs at a social level. It is in *The Elementary Forms of the Religious Life* that Durkheim famously, explicitly, refers to time. Durkheim attributes all experiences of the world to a set of categories, a legacy of Aristotelian philosophy[11] (Durkheim 1995, 8). One of these categories, time, cannot be conceived of except in periods, whereby trying to comprehend "a time that was not a succession of years, months, weeks, days and hours . . . would be nearly impossible" (1995, 9). These conventional time periods are socially derived, meaning "what the category of time expresses is the time common to the group, a social time" (Durkheim 1915, 11). The category of time is in this view an "objective reality" for Durkheim, in terms of how one actually lives. Time indeed only manifests through, or is produced and maintained by, collectively reiterated human practices. This is a rhythmic reproduction that is socially alive before new individuals reanimate it, and which "corresponds to the periodical recurrence of rites, festivals, and public ceremonies at regular intervals . . . taken from social life" (1915, 9). The earlier accounts of how societies each structure lateness differently would not therefore be surprising for Durkheim, given that according to his reading, time is only ever socially derived. This plurality does not prevent each social derivation from being an "objective" and factual form of time for those living it though. What is most noteworthy at this point about Durkheim's appraisal, therefore, is this characterization of the "objectivity" of temporal rhythms produced by the human, social realm. Preceding arguments that we have covered have demanded that what seems objectively regular about a social time structure is conditioned by a source that is celestially separate from that social realm. Durkheim instead attributes the source of what is objectively regular about a social time structure to what is socially inside time's construction.

Via sociological and anthropological accounts of global lateness protocols, differing regional, historically normalized, socially derived time periods become apparent. Individuals take for granted and adhere to the localized social time structure after being born into it. This is part of Durkheim's point. Time does not represent a separate reality, because the category of time only has a social reality. When an individual is late, they do not think of their lateness as being conditioned by the contingent, social construction, of what is actually celestially separate about time. Rather, they live by what is collectively instituted as time, and if they breach accepted protocols, will expect to face the resulting social punishment. Recognizing Durkheim's emphasis on this socially internal source of social time's objective regularity serves to differentiate his structuralism somewhat from the social constructionist time condition.

Nevertheless, it is as a result of incorporating Durkheimian structuralism that a question emerges regarding whether Durkheim replaces the constructionist sequence of an objectively separate, real-world time, which precedes myriad social times, with a new sequence in which the social structure of time externally precedes individual time(s). If individuals are born into an already existing social structure of time, is there still a transcendent relation between time's objective source and their subjective versions of it? This would matter, given that the singularity of an "objective" temporalizing, separated from the multiplicity of different times of individuals living within it, would simply maintain or reproduce the constructionist opposition between an objectively real time, and the myriad constructed times coerced by a preceding force.

FROM OBJECTIVELY EXTERNAL RHYTHMS, TO SUBJECTIVE RHYTHMS

It is not only that time is socially, objectively conditioned, for Durkheim. A society's continuity of behaviors are said to be produced by a network of influences of which individuals will often be unaware. Even if an individual's acts appear to express personal motives and desires, by conforming, as Durkheim states, to "my own sentiments and I feel their reality subjectively" (Durkheim 1938, 1), such reality is still objective. This reality "does not cease to be objective" (1938, 1) because the associated behavior represents something factual, established, and actual about that social structure. Durkheim's use of the term "objective" can be linked to his conception of "social facts." Social facts are introduced here in *The Rules of Sociological Method* as objectively[12] collective, coercively durable ("objective, resistant, and persistent"), ways of socially being. These socialized orientations are differentiated from the individualities of subjective consciousness (xiii).

Importantly, a "social fact" is not simply the sum of individual manifestations. If the social fact, which is a single collective, was the average of the aggregate of its individual manifestations, this would imply a divisible entity, giving the individual an autonomy not present in the objective, collective, social structures that he observes. Social phenomena are different from each individual consciousness for Durkheim, meaning that social "facts reside exclusively in the very society itself which produces them, and not in its parts, i.e., its members" (Durkheim 1938, xlviii). In tracing individual thoughts and actions to objective, collective, social forces, Durkheim questions, as we have noted, whether suicide is an autonomous, self-motivated act. By analyzing the statistical distribution of suicide rates, uniform patterns illuminate for Durkheim the socialized compulsions that determine such consistency, where the "regularity of statistical data . . . implies the existence of collective tendencies exterior to the individual" (Durkheim 1952, 283). As indicated, these collective tendencies are indeed considered to be responsible for the durability of a number of social trends, Durkheim asking not just what mysteriously causes humans to suicide uniformly but what regulates the rhythms of any social practice/behavior (1952, 270).

The objectivity of this coerced regulation, in predating and transcending all individuals, means that it is the source from which all behaviors, normative or non-normative, legal or criminal, manifest (Durkheim 1938, 1–8). According to such logic, I suggest that being late can be conceived not as that which is caused by an individual's capacity to break from social coercion. If all behaviors are socially objectively shaped, then the promptness as well as the tardiness of individuals from the same society is generated by common mechanisms. This position also contradicts interpretations that non-normative body modification practices represent deviant breaks from society. Such an interpretation would perceivably be problematic for Durkheim, given that according to the objectivity of the social forces that he identifies, it would have to be argued that the motivation to modify the body in a non-normative way manifests from the same source that engenders normative body practices. Non-normativity is irreducibly entangled with normativity, whereby the body modifier who creates a non-normative corporeality is equally bound up in the production of normative body frames. Social practices such as crime, suicide, lateness, and body modification, are living, productive, individual manifestations of a collective field from which nothing is excluded.

I note two curiosities here. First, that behaviors that might typically be interpreted as personalized indicators of an individual's identity are seen to represent something significantly impersonal. Furthermore, that secondly, behaviors such as lateness and suicide are counterintuitively presented as evidence of a social uniformity and cohesion, rather than of a society's dysfunction. The curiosity of this ramification that arises from a structuralist perspective is compelling. The act of being late, which should compromise the synchronization of society, can potentially be recognized as arising from a synchronous and unified source. Perhaps even more provocative is the insight that for Durkheim, an individual being impelled to suicide is indicative of a regulated, social consciousness rather than of that which occurs uncontrollably.

According to the position of social constructionism with which we are engaged, for a rhythm to be uniform and objective, its source arrives from a pre-social, pre-representational level. Conversely, Durkheim posits that the source of an objective rhythm can be immanent within the human, social world. Despite this difference, in a certain regard Durkheim's argument is considerably social constructionist in that it argues that the experience of being human is completely governed by a socially produced, periodized reality. There is nothing but socially factual phenomena. In another regard, however, the feature of social constructionist thought that posits a subsequent social representation of an already existing, celestially external, natural reality does not fit Durkheim's outlook. As we have seen, for Durkheim there is not a reality to which to refer that is outside the socially derived categories.

At the end of the last section we noted though, that where social constructionism externalizes the objectively real phenomenon from each subject's experience of it, so Durkheim characterizes the objectivity of collective force as external to each individual experience and reproduction of social facticity. In *The Rules of Sociological Method*, Durkheim introduces social facts as being "noteworthy" because they are ways of acting, thinking and feeling that exist *outside* the individual consciousness (Durkheim 1938, 2). This grounds his claim that individual behaviors are not reducible to autonomous subjectivities. Indeed, the objectivity of the social sciences for Durkheim manifests via this externality of social forces from individual subjects. This reproduces the objectivity of the "natural sciences," which is attributed to the separation of an object-thing from any subject-consciousness. Durkheim duly states that just as "idealists separate the psychological from the biological realm, so we separate the psychological from the social" (1938, xxxix). It is not that social facts *are* material objects/things for Durkheim, but just that they should be treated in the same right (xlii). The relation between the subject and the social-fact-as-thing should be viewed in the same way as the physical sciences (and for our interests, social constructionisms) conceive the relation between the subject and the real-material-thing. The objectivity of a "thing," beyond subjective idiosyncrasy, for Durkheim assures its regularity and examinability beyond individual contingency (28).

One interrogation of the externality by which Durkheim characterizes social phenomena/facts that I would like to suggest, emerges in the observation that while the social fact is exterior to the individual, it is fabricated by a social collective of which each individual is a constituent. Could this be read in a manner in which each individual is in fact, *in the fact*? Regarding social facts, Durkheim does explicitly state that "we have collaborated in their genesis" (Durkheim 1938, xlv). While in terms of their (re-)production, "each one of us participates in them only as an infinitesimal unit" (1938, xlv), social facts still reflect the diversity of individualities. There is no coercion of individuals without individual influence. Of this, Durkheim is adamant; "there is no conformity to social convention that does not comprise an entire range of individual shades" (lvii).

It is here, in the preface to the second edition of *The Rules of Sociological Method*, that Durkheim indeed refutes readings of the first edition that characterize social facts as deterministic (Durkheim 1938, liii). By again referring to the physical sciences, Durkheim notes that "every physical milieu exercises pre-constraint on the beings which are subject to its action," manifesting with "pre-determined effects" (1938, lv). However, there is a difference, he instructs, between the "modes of coercion" that occur in the social milieu versus the "rigidity of certain molecular arrangements" of the physical milieu. Social facts are conversely malleable, being shaped by whichever institutionalized forces are prevalent. The collectivity of any era's social-fact-as-thing manifests uniquely, via a collaboration in which "several individuals must have contributed their action; and in this joint activity is the origin of a new fact" (xlv). Nevertheless, it should not be forgotten that this collaboration has an *external* development for Durkheim, in that this "joint activity takes place outside each one of us" (lvi).

With Durkheim's distinction of physical and social phenomena, a singularly predetermined objectivity of the natural realm is distinguished from the myriad objectivities of different social structures. The social constructionist time condition's separation of a singular natural temporality, from plural social temporalities, is hence apparent. Nevertheless, while the individual is positioned outside social objectivity, the limit between them, between the subjective and the objective, is blurred. I intend to interrogate this limit in two ways in the next chapter, in order to further identify and establish what is at stake in the social constructionist time condition. Firstly, we will ask whether Durkheim's insistence that the social fact is exterior to the individual is conditioned by the assumption of an absolutely *prior existence* of what is objectively social. Secondly, individuals are recognized by Durkheim as being co-constitutive of social phenomena. Yet the notion that social facts are "outside" each individual consciousness conversely, and contradictorily, demarcates individuals. Indeed, such division seemingly conflicts with the inescapably socialized subjectivities that manifest via Durkheim's structuralism. In response, we will investigate whether the individual-social relation is as separate as this model demands, by considering the role of subjective time-consciousness. Are humans really outside what for this sociological theory is the objective reality of time?

Chapter Three

Subjective Times

Transcending the Present?

> Time—our youth—it never really goes, does it? It is all held in our minds.
> —Helen Hooven Santmyer, *She Said What?* (Cannon 2012, 130)

AUGUSTINIAN AND ARISTOTELIAN PRESENTS

This chapter marks the final stage of the process of identifying the condition that is implicit to the social constructionist perspective on time. A specific agenda in this chapter will be to interrogate the position, lingering from the previous chapter, which separates time's objective source from subjective times. This continues how, in the time theories reviewed, we have encountered a series of separatisms based on presence|absence dichotomies. Time's reality is presumed to be present in naturally celestial time, but not in the myriad socialized representations of it. Similarly as we have just seen, an objective consciousness, in which the category of time is present, is positioned outside the socially constructed, subjective, impressions of time. Real time is seemingly never present for the individual in either view, but rather operates at the kind of distance that has emerged as a hallmark of the social constructionist time condition.

Given this concern regarding the presence of real time, we will commence this chapter's inquiry into subjective times by observing some of philosophy's formative considerations on time and presence. Such retrospection is a relevant move in terms of the critical features of our study, adhering as it does to the originary characterizations of critical theory as enacted by the Frankfurt School. This concerns philosophy's self-reflection on the development of its entrenched ideas or perspectives, as well as its capacity to intervene and dismantle such perspectives where necessary.

To this end I first turn our attention to the musings on time of Saint Augustine of Hippo. Indeed, in acknowledging Augustine's relevance to any inquiry into time, Barbara Adam notes that while "not every treatise of time refers to Heraclitus, Kant, Hegel, Husserl, Bergson, Heidegger, Schütz, or Whitehead, I have not come across a single study that does not mention St. Augustine's *Confessions*" (Adam 1990, 33). Augustine meditates upon the nature of time in *Confessions* (1961 [circa 400]), observing that it is only conceivable as past, present, and future. While the past no longer is, and the future is yet to occur, the present-instant is said to have neither duration nor extension, otherwise it would pass into the nonexistence of the past (Augustine 1961, 266). However, if only something in existence can be extended, whereby "something must exist to be capable of being long" (1961, 266), the question is prompted about where is temporal extension? Augustine duly surmises that past and future do exist, but are hidden, only emerging when passing as the present (267). This "hidden" characterization of time's states, entirely dependent on the overt present, evokes the conceptual polarization of material space from immaterial time that was earlier encountered. Material things in the spatial/physical world are characterized as tangibly real and accountable, they are evidently present. Conversely, time is mysterious and invisible. It lurks, as Elizabeth Grosz describes, as "a silent accompaniment, a shadowy implication" (Grosz 1999, 1).

Potentially countering the unaccountability of time's invisibility, we also see that time is culturally spatialized. Not for the last time in this book, I posit that when considering lateness, the spatial representation of time on a clock, or via "natural" markers such as the sun's physical position in the sky, demands that an individual's lateness has nowhere to hide. In fact, the celestial realm of all physical, planetary bodies, has a temporality that is readily tracked through spatial movement. We have seen that African, Caribbean, and Filipino time structures have been characterized as "present-centric" due to their supposedly "closer alignment" with the material and spatial constitutions of ecological and celestial rhythms. However, in all global regions it could be said that time is regulated by something spatially and presently visible about its forms, such as is identifiable on clocks and calendars. These themes will return with prominence in chapter 6.

In order to explore time's potentially hidden existence beyond the present though, Augustine discusses one's awareness, or consciousness, of extended time. Because the present is always transitioning to nonexistence, time's extension for Augustine can only be measured in the present, "while it is passing, for no one can measure it either when it is past and no longer exists, or when it is future and does not yet exist" (Augustine 1961, 266). Here I would like to ask whether this conception of a hidden past and future is similar to the temporality of social rhythms in Durkheim's theory. For Durkheim, individual behaviors are not merely a present phenomenon. Rather, they reflect how the present period has been informed by preceding periods, evidencing "in the past the germs of new life which it contained" (Durkheim 1952, 359). The social past lurks in the social present as a trace, just as for Augustine the subjective past exists by continuing in the subjective present-as-passing. Where for Augustine, the future exists only as an anticipation of its manifestation as a present-that-will-be, similarly for Durkheim, the future is only possible because its subsequent potential already lurks in, or is "contained in" (1952, 332), present uniformity.

If we reflect on what might have informed a theory's present-centric tendencies, a tradition needs to be recognized that includes Aristotelian philosophy. Aristotle's position on the presence of time derives from the influence of Plato's *Timaeus*, in which time is presented as a "moving image of eternity" (Plato 2008, 37d). Plato corresponds time with a certain kind of universal motion, exemplified in the "stars which were necessary to the creation of time and had attained a motion suitable to them" (2008, 38e). Of particular importance is that in describing time as this "wandering of bodies," time must be seen just *as* the celestial motions, rather than as the human measurement of them (39d).

Aristotle develops this conception in his *Physics* by arguing that time and motion correspond but are not identical. This is because time is everywhere, whereas motion is in particular things (Aristotle 1996, 4.10.218b5–4.11.219b9). The Aristotelian perspective here is that motion is an indication of a thing's displacement, or of how we recognize its change in position from a "starting-point to an end-point" (1996, 4.11.219a10). Time is not identical to such change; however motion or change is an indication of time's existence (4.11.219a10). That the before-and-after points of this change are appreciated by number according to Aristotle develops the Platonic theory of time and motion into a recognition of the magnitudinal character of time, in that "change follows magnitude, and time follows change" (4.11.219b16).

In numbering the change between a thing's before and after states, Aristotle argues that what are countable are instants or "nows." It is through the "now" that we will get a sense of the presentism of Aristotelian time. The "counting" of the nows does not indicate their aggregation, but rather the marking of *qualitatively differentiated* before-and-after end points that are "successively different" from each other (Aristotle 1996, 4.10.218a8). This successive character does not mean that such nows are merely lined up next to each other. The reason for this is that nows do not simultaneously exist to be situated side by side. As subsequent nows manifest, earlier nows cease to exist, meaning that for Aristotle, "nows will not be simultaneous with one another" (1996, 4.10.218a8).

We are exploring this constitution of nows because of the role it plays in Aristotle's conception of the present. For Aristotle, a crucial question is whether a subsequently present "now" either remains identically the same as it in turn is replaced by another now, or alternatively is always becoming different. His response, discounting the possibility of simultaneous, side-by-side nows, is that the past now, in being superseded by a subsequently "present now," does not entirely cease to be present itself, but still participates to a certain degree in the "subsequent" now in which it is "contained" (Aristotle 1996, 4.10.218a8). Each now according to this perspective maintains a certain, perpetual presence, retaining its identity, while also always becoming a qualitatively different now given its implication *in* the nows' succession (1996, 4.11.219b26). This marks the paradox of continuous Aristotelian time. A now is the definable condition of the before and after in change, "determining time" given that "in so far as it is to be found at successively different points, it is different—this is what it is to be now" (4.11.219b12). Complementarily though, the now is *not time* as a number of change but in preserving its identity remains a durationless, present instant, as "something single and identical" (4.11.219b12).[1]

The point we should take from this is that the Aristotelian, durationless, present instant is not actually *in time*, a feature that philosophy has since critiqued. The most prominent interrogation is perhaps offered by Martin Heidegger, who in *Being and Time* (2010 [1927]) argues that in Aristotle's *Physics* "time shows itself for the vulgar understanding as a succession of constantly 'present' ['vorhanden'] nows that pass away and arrive at the same time" (Heiddeger 2010, 401). It is Heidegger's impression that this "vulgar" hierarchization of the now's presence is "no accident" (2010, 401), but in fact is identifiable in philosophy's history of its inquiries into time (410). Heidegger laments accordingly that a spectrum of fields within philosophy, as well as understandings in everyday discourses, have adopted the presentism of Aristotelian time that "keeps itself solely in the perspective of commonsense" by presuming a world that is permanently present at hand (401). Not only is the "now" misinterpreted, Heidegger asserts, to be invested with a kind of pure presence outside temporal flow, but also "the succession of nows is interpreted as something somehow objectively present" (402).

A further attack on the philosophical heritage of the presence of the now is launched by Heidegger in *Kant and the Problem of Metaphysics* (1962 [1929]). Here Heidegger targets the Aristotelian characterization of Being as "synonymous with *permanence in presence*" (Heidegger 1962, 249; author's original emphasis). According to Heidegger's reading, Aristotle makes the mistake of interpreting that Being is entirely apparent at any given, isolated point. This is the aforementioned durationless instant that determines the "Being" of time from the "point of view of the *now*, that is, from the character of time which in itself is constantly present and, hence (in the ancient sense of the term) really *is*" (1962, 250; author's original emphasis). As scholars of Heideggerian philosophy will be aware, and as we will indeed touch upon in chapter 5, Heidegger's alternate conception offers an awareness of time that operates via the unified function of three "ecstasies": the past, the present, and the future. It is from this kind of imperative that Jacques Derrida will later reflect on, and further critique, philosophy's perceived reliance on presumptions of present-centrism.[2]

The reconfiguration of dominant notions of "permanence in presence" can indeed be identified as an historically critical maneuver, if we integrate perspectives from the so-called "second wave" of critical theory offered by Jürgen Habermas.[3] In *The Structural Transformation of the Public Sphere* (1991 [1962]), Habermas reviews the development, and the demise, of the bourgeois public sphere in the eighteenth and nineteenth centuries. Habermas particularly wishes to interrogate the supposedly "free" institutions of contemporary society, evidenced by his commentary around the term "public" (Habermas 1991, 1–2).

In tracing a history of impressions of what constitutes the "public" arena, Habermas describes the transition from older eras to the aforementioned, modern bourgeois society. Older cultures he characterizes as "representational." This refers to the necessity within such environments to symbolize the authority of those in power in a publicly visible manner, whereby "the attributes of lordship . . . were called public . . . for lordship was something publicly represented" (Habermas 1991, 7). Habermas states that this sense of "public" does not equate with what we might consider to be constituted by a "social realm." Instead, public is said to indicate the lord's presentation of a "status attribute" (1991, 7).

It is in the "presenting" of what is public as a sense of status or power that we might begin to see the direct relevance of this incursion from Habermas, to the current theme of present-centrism. Habermas believes the way that power presides in these past, representational modes evokes the durationless qualities of a timeless, eternally present, condition describing how an authority figure "presented himself as an embodiment of some sort of 'higher' power" (Habermas 1991, 7). It is as a result that the church is reported by Habermas as being one of the "last pillars of representation" (1991, 8). Authority in this interpretation presents with an in-itself, essential quality, the "really is" or the "permanent presence" of Heidegger's Aristotelian commentary that we have just reviewed. The institutionalized power of authority figures is guaranteed, because it is represented as ever-present, Habermas describing a mechanics in which the "nobleman was authority inasmuch as he made it *present*" (13; my emphasis).

The representational presentism is eventually superseded according to Habermas through the genesis of capitalism. With capitalism comes an increased exchange of products, services, and information, which particularly in the eighteenth century includes "the traffic in news that developed alongside the traffic in commodities" (Habermas 1991, 16). From this generation of information, a different form of "public" space is said to develop beyond that of authoritarian symbols. As new periodicals and journals circulate in which "private people" are able to communicate opinions and information, authority is made to legitimate itself before public opinion (1991, 25).

Habermas links this dispersal of ideas to the physical environments of new public spaces, such as coffee houses, in which such publications could be readily found (Habermas 1991, 31–32). These arenas, and the discussions generated within them, are portrayed as critiquing the previously taken for granted presence of authority. It is as a result that Habermas characterizes this kind of discourse, and indeed the era in which it was facilitated, as "critical." Such critique is reported to cover many areas of modern society, described as "literary at first, then also political" (1991, 32). While the prior, representational culture, which projected a pure or timeless presence of authority, produced a docile audience, the critical contributions found in these new arenas questioned the presently dominating institutions in all aspects of society.

The characterization as "critical" of an interrogation of an authority that had been interpreted to be ever-present informs another way in which our inquiry into the social constructionist time condition can also be viewed to be a critical endeavor. This concerns how we question accounts that reduce African, Caribbean, and Filipino lived experiences of time to a timelessly present-centric, non-abstract, natural power. Presumptions of the authority of nature-as-present lurk within the conceptual homogenization of something like "African time," portraying a time structure that is said to be bound to a durationless moment that lacks an awareness or construction of an abstract future. Such a view does not absolutely suggest that African time cannot frame developmental processes, but rather demands that the African temporal experience is, as Anthony Aveni observes, a series of isolated, presents (nows) that "place no value on the future" (Aveni 1989, 332).

A related point becomes apparent here that we have touched on from Pierre Bourdieu, who explains that in such a context, while "various outcomes are always possible" from any particular present, the future is not even acknowledged as anything but an expected present (Bourdieu 1963, 62). For the Algerian population featured in Bourdieu's work, this expected subsequent state is a present that simply does not exist yet. There is in such conceptions no accommodation of a linearly progressive series of simultaneous states, but rather a succession of Aristotelian durationless presents. We should remind ourselves that for our interests, the homogenized "African time" concept has not been integrated as an example of a foil to the concept of Western time, nor as a category whose individual components will be unpicked in this work. Rather, it illustrates how socialized time is conceived to manifest in multiple forms globally, certain of which are problematically theoretically positioned as being more present-centric, and closer to the possibly timeless authority of natural time, than other cultures' socialized times.

We can also revisit the insight that time structures that heavily regulate social lateness are conversely posited as sharing an intimate relation with the abstract linearity of time. Where the future can be anticipated and symbolized, so the interpersonal synchronization concerning it can fail, and late behavior can be adjudicated accordingly. The result is that present-centric social time structures with apparently less strict lateness protocols are conceptually distinguished, as we have seen, from those cultures whose abstractions of time have supposedly distanced them from a natural state.

Taking into account the interpretation that a present-centric temporality or rhythm is a more originary form of time than its myriad, abstracted, complicated forms, we can revisit the timing of Durkheim's objectively social temporality. Durkheim posits a social structure that is authoritatively present as time. Social time is not perceived even as a representation of time, it *is* time. Does this mean, therefore, as we have indicated we need to ask, that the social for Durkheim is separately prior, as *a permanently established past-presence*, to the externalized subjects who temporally comprise it. As we begin in this chapter to open philosophical perspectives that can speak to the condition that excludes socially constructed, subjective time consciousnesses, from time's real/actual source, so George Herbert Mead's philosophy of the present becomes relevant.

MEAD'S METHOD PRESENTS: THE NEVER ISOLATED PAST

In *The Philosophy of the Present* (2002 [1932]), Mead acknowledges, similarly to Augustine, the assumption that past and future are what has been, or are what is yet to be, and to "both we deny existence" (Mead 2002, 35). In suspecting there is more to time than this, however, for Mead such an inquiry should not be concerned with the past when it was previously present (2002, 46). This is because when that past was present it did not have the status of being past.

Conversely, Mead is interested in the *relation* of the past to the present. On the one hand, Mead describes this relation as irrevocable (Mead 2002, 36). This coheres with the exclusively forward-moving time-linearity of pasts conditioning present realities apparent in Durkheimian theory. Furthermore, it is consistent with the commonplace interpretation that time has one direction and that direction is forward-moving from past to present. Mead defines this irrevocability as "the necessity with which what has just happened conditioning what is emerging in the future" (2002, 47). Congruently,

Durkheim describes such progression as the way past social incarnations inform subsequent, present socials, where our "social institutions" were "bequeathed to us by former generations" (Durkheim 1938, xlv).

On the other hand, Mead also characterizes the past as "revocable" (Mead 2002, 36). My reading is that this counterintuitive notion of a past that is revocable can be explained according to the novelty of the new present. With the manifestation of the present, the past that conditions this present must also become something that it was not before this relation. This novel relation with a now manifested present reconditions the conditioning past as a "different past" (2002, 36), because of the past becoming involved in a newly presented production that it had not previously constituted, nor by which it had been constituted. This new relation reconstitutes the past as that which will have become *this present's* past. Mead succinctly describes of this, that it is the "what it was" of that past "that changes" (37), referring to the localized particularity of any past-present construction:

> The past is there conditioning the present and its passage into the future, but in the organization of tendencies embodied in one individual there may be an emergent which gives to these tendencies a structure which only belongs to the situation of that individual. The tendencies coming from past passage, and from the conditioning that is inherent in passage, become different influences when they have taken on this organized structure of tendencies. (Mead 2002, 48)

This argument can illuminate our considerations of the temporality between the objective social structure and the individual in Durkheimian theory. I recognize three ramifications that arise from Mead's passage to this extent. Firstly, any individual behavior, such as suicide, can be read as an "organized structure of tendencies" that, when emerging in the present, (re-)incarnates the past social tendencies as the aforementioned "different influences," in a new, co-implicated with/as the present manner.

Secondly, I argue that each suicide participates in the production of a social rate/rhythm of suicide as a continuation, but not as an identical replication, of a preceding rate/rhythm. Indeed, the regularity of the past suicide rate/rhythm only manifests in the present, whereby the "what it was" of the past becomes not simply that which *has* conditioned the present suicide rate/rhythm, but also that which *will have* been conditioned by the manifestation of the present suicide rate/rhythm. Or in Mead's terms, in relation to the past, the present represents "the future [that] is continually qualifying the past in the present" (Mead 2002, 65).

Thirdly, the present individual participates in the production of the social at the same time as they are produced by the social. This concurrent manifestation of cause and/as effect is explained by the distinguishability of an effect of past social forces (a suicide), which simultaneously emerges as a cause in maintaining the uniformity of social rates/rhythms of suicide. From this I believe that the co-production of social and subject means that the social-as-past-conditioner-of-the-present does not simply preexist the individual-as-present-expression. The assumption of this sequence is an issue that has emerged with Durkheim's sociology, and indeed is something that I have claimed mimics the social constructionist sequential separation of natural and social phenomena. Instead, what is now being recognized is that the present-individual-expression (re-)produces the social-as-past-conditioner-of-the-present, affirming Mead's notion of a past becoming what it was in the present.

This different philosophy of the relation between the past and the present provides a basis upon which to critique Durkheim's sociological notion that social facts are objective because they preexist the subjects that they coercively produce, and that, as Durkheim states, "their existence prior to his own implies their existence outside of himself" (Durkheim 1938, 2). Durkheim's sociology fixes the past permanently in place. Past social structures are productive of present individuals, but are not produced by present individuals, given that these latter individuals "took no part in their formation" (1938, xlv). Conversely, Mead's conception of time supports my reading that what actually occurs in Durkheim's model is not only a past-presence producing/authorizing the present, but also a present participating in producing/authorizing that past.

Durkheim's theory installs a forward-moving sequence. Socially constructed worlds precede subsequent generations of humans fashioned by these collective, socially constructive, templates. What is now being considered, though, is that the transition from social structure to individual experience is not straightforwardly sequential. The continuously regular, but not identical, relation of current suicide rates to those of a preceding period produces the simultaneous realities of conditioning-and-conditioned past-and-present societies and human tempos. While the primacy of the manifestations of the present is duly acknowledged in this version of time's social construction, such a present is not outside time as per the Aristotelian now, nor does it restrict, as seen in portrayals of "African time," experiences of time to a durationless, present state.

Following Mead's position regarding the counterintuitive direction and production of temporal states, I anticipate that certain readers might take the contrary view that it is merely our impression of the past that has changed. Complementary, such readers could hold that the "past in-itself" is still, permanently, as it was prior to the arrival of a new present. This sense of a permanently established past-presence, however, is the type of reading that Mead can complicate. Inspired by his interpretation, my assertion is that a past "in-itself" is not a "past," nor is it even a separate state of time/temporality. Rather, it is only via that past's co-relation to the present that the past, as time/temporality, becomes the past. Mead describes past-present co-production generally as a perpetual state, in that "we are not contemplating an ultimate unchangeable past that may be spread behind us in its entirety subject to no further change" (Mead 2002, 57). For my application to Durkheim's sociology of this logic, if Durkheim's social facts are therefore objective, it is not because they transcendently preexist and authoritatively present human tempos. What is instead objectively occurring is a past-source of human temporality that is always inescapably malleable *with* human presence.

We have seen the social constructionist position under examination in this book posit that our constructed version of the world never accesses the already established reality that conditions it. However, the point I would like us to take from Mead's philosophy of time is that if any time state is perpetually reproduced, this problematizes the assumption that the source of anything is simply an origin that is eternally fixed in the past, away from which new presents are increasingly distanced. If present constructions of time, and a supposedly separately preceding real-source of time, were to manifest via the Meadian conception of interdependent temporal relations, the notion of "access" to an originary, distant time-source, could be contested. This begins to suggest possible porosities between the social constructionist time condition's naturally antecedent, and socially subsequent, temporalities. More immediately though, in blurring what was broached to be the correlative separation between the timings of the social, and of human subjects, the question now presents; at which point is something social? What indeed is sociality?

Mead's response to this question exhibits an important consistency with our current considerations of the simultaneous production of states of time. According to Mead, emergent phenomena are concurrently in both the new (present) and the old (past) systems. This systemic plurality defines, for Mead, socialization. Sociality manifests because "in the passage from the past into the future the present object is both the old and the new" (Mead 2002, 76–77). My interpretation here is that Mead is evoking the usual equation of sociality with *inter*subjectivity, but attributing such *inter*activity to already co-constitutive, *systemic* states. Sociality emerges via systemically plural present-and/as-past, where "the social character of the universe we find in the situation in which the novel event is in both the old order and the new which its advent heralds" (2002, 75). Such plurality, I believe, can inform an understanding of the sociality involved in Durkheim's "mysterious" connection between the subject, and the temporality of objective, social rhythms.

CAUSATIVE, AGENTIVE SUBJECTS: STRUCTURALIST INTERSECTIONS

The individual's involvement in the regularity of socially objective rhythms is the "mystery" to which Durkheim refers. Durkheim's sociology asks how an individual can inform or infect how another individual from a different time period enacts suicidal tempos and rates (Durkheim 1952, 273). That individuals do perpetuate this trans-temporal behavioral regularity, for Durkheim redefines assumptions concerning suicidal causality. The coercion of individuals to kill themselves represents a condensation of the personal and the impersonal, contradicting closed equations restricting suicidal motivation to sovereign, individual misfortune/sadness.

As earlier noted, a dispersed sense of the authorship of one's actions seemingly contradicts liberal notions of power and choice, such as exhibited by body modification practitioners who commonly define their practices in terms of a sovereign control *over* corporeality. Complementing this mentality are often found assertions that a practitioner is contesting, and operating outside, normal social regulations. This is consistent with the impression of autonomous suicidal causality, against which Durkheim is arguing. Here the self-curating thematic of body modification discourses characterizes the body as a personal self-projection, presuming, Pitts observes, that "such practices exteriorize an 'inward depth'" (Pitts 2003, 31). Similarly, Mike Featherstone notes the common motivation of such practices as "taking control over one's body, of making a gesture against the body natural" (Featherstone 2000, 2). This sense of corporeal control permeates the body modification community, reflected in its manifesto found at the aforementioned *Body Modification Ezine (BME)*. BME's founder, Shannon Larratt, duly states that "a person has ownership of their own body, and we don't have the right to try to take away that sovereignty" (Larratt 2004).

By assuming an individualistic self-creation, social forces are kept at a conceptual distance, particularly where non-normative practices are concerned. Some women, for example, describe genital piercing as a way to rebel against patriarchal dominance and to "reclaim" power over their own bodies (Pitts 2003, 3). Such individualized body "recovery" envisages a demarcation from the social, polarizing the individual and the social combatively. As Eric Gans observes, what is presumed here is that body modifiers are able to "designate their bodies as loci of resistance" (Gans 2000, 165). Conversely, our investigations to this point have not characterized such resistance as originating outside or against the social but rather as expressing the social's collective, objective, yet differentiated forms of coercion.

By weaving through the theories of Mead and Durkheim, I duly adopt a stance that generally accords with the well-known position of Michel Foucault (1977), which argues that subjects and bodies are never outside, nor subsequent to, power. Neither, it must be suggested for a Durkheimian context, are subjects simply produced by *preexisting* social structures. Rather, what I identify in terms of body modification practices is that such bodies manifest as embodied insights into the myriad social forces from which one is genealogically produced, and which one participates in producing. Whether in terms of bodily, or suicidal, tempos, social facts produce individuals as objects who concurrently produce such objective facticity.

This characterization of individual objectification is important, given that when the incarnation of the subject is implicitly involved in the production of objectively subject-shaping social structures, cause and effect manifest concurrently. The distinguishability of an effect, such as a suicide, only manifests via its simultaneous emergence as a cause that participates in the uniform, social rate of suicide. Furthermore, if we adopt Mead's characterization of time, the emergence of the "present," whether as a social or individual state, produces, and is produced by, concurrent pasts and futures. The body modifier is not simply an individual whose motivation to tattoo their face or undertake plastic surgery is secondarily reactionary to preexisting social forces and temporalities. We are not encountering an oppressive social construction of the subject by what is objectively separately real about human tempos. Neither, conversely, are such subjects indicative of a defiantly demarcated self-production. Rather, the body modifier's subjectivity is perpetually bound up with a broader social objectivity.

This is a significant ramification for the temporality of Durkheim's social facts; the simultaneous incarnation of subject and object. According to the integration of Mead's perspective, a co-productive relationship can be identified between present individuals and established social structures. This is sociality. There are related questions that we can ask from this development concerning the context of social lateness. If individuals and social structures are co-constitutive, does this mean that present individuals therefore have an agentive role to play in the social constructions of time that pervade their cultural or collective context? Furthermore, if the answer to this question is yes, how does an individual's tardiness, or asynchrony in a particular cultural setting, affect that culture's collective notions of lateness?

The argument, as I have developed it, is that individuals are social constituents, producing, rather than merely following, past social incarnations. Recognizing this facilitates an engagement with the renowned critique from sociologist Anthony Giddens that Durkheim installs a constrained individual who is devoid of agency. In *The Constitution of Society: Outline of the Theory of Structuration* (1984), Giddens demands that structural sociologies suffocate subjects by imposing "circumstances, of which agents are ignorant and which effectively 'act' on them, independent of whatever the agents may believe they are up to" (Giddens 1984, xix). Giddens sees Durkheim's study of suicide as a prime example. Conversely, Giddens does not want to discount the role of the individual, in that "there are some acts which cannot occur unless the agent intends them. Suicide is a case in point. Durkheim's conceptual efforts to the contrary" (1984, 8).

I introduce Giddens to this discussion because his issue with Durkheim seems to cohere with my concern about Durkheim's use of the terms "exteriority" and "externality" to describe the subject's relation to social forces. For Giddens, the defining quality of Durkheimian structural properties is that they are "certainly exterior to the activities of the 'individual'" (Giddens 1984, 170). Giddens critiques this thesis, arguing that "structure is not 'external' to individuals," but instead "as memory traces, and as instantiated in social practices, it is more 'internal' than exterior to their activities in a Durkheimian sense" (1984, 25). While Giddens and I cohere on this point, I distinguish my concerns from Giddens's in recognizing that I believe Durkheim uses these loaded terms to emphasize that behaviors (such as suicide), and consequent human tempos, are not exclusively motivated by individual, subjective, agentive causation. Given my reading of individual-social co-constitution even in Durkheim's sociology, I therefore argue that Durkheimian social structure manifests not as pure external constraint as Giddens supposes but rather as an expression of a conditioning/conditioned individual. Giddens and I actually agree that "structure is not to be equated with constraint but is always both constraining and enabling" (25). Our interpretations differ though on the nature of the constraint and exteriority in Durkheimian sociology. As noted in the last chapter, Durkheim directly refuted these kinds of accusations that his structuralism overly constrained the individual.

Justification for how I am interpreting Durkheim can be attributed to the conception of time developed with Mead's considerable assistance, in which a past-presence is not permanently established. The interpretation of a permanently antecedent social structure is reconfigured accordingly. This alternate perspective could even be said to be evoked by Durkheim himself, via his aforementioned discussion of the "individual shades" that are involved in the "genesis" of the social fact. Contrary to Giddens's reading of Durkheim, we can assert that it is not purely a "constraint [that] stems from the 'objective' existence of structural properties that the individual agent is unable to change" (Giddens 1984, 176). Such a definition relies upon a unidirectionally linear, cause-effect chain, between a preexisting social and the subsequent "will" of an individual. Conversely, I have argued that what is objective about Durkheimian social rhythm is the constitutive participation of all individual agents. Individuals do not will change to an already existing social. The individual instead is always already inescapably participating in/as social (re-)construction.

These insights can be used to frame responses to the earlier questions concerned with the role of the individual in collective understandings of lateness. This speaks directly to the context presented in the introduction of this book regarding the politics by which lateness differentially manifests globally. If the individual is interpreted, as I believe they should be, to unavoidably participate in social structuration, then it is not that an individual, born into a context such as the Philippines and frustrated with the supposed perpetuation of tardiness, separately or defiantly *wills* different regulations regarding lateness. Inspired by Mead's philosophy of time, I want to argue that what is socially structured in terms of lateness protocols does not represent a past that temporally *transcends* an individual's present reaction against it.

Individual-social co-constitution has in this chapter instead marked the *immanent* co-constitution of human pasts, presents, and futures. Conversely, the influential theory of Barbara Adam characterizes these time relations in *transcendent* terms, supposing that "human time is characterised by transcendence. . . . All human action, for example, is embedded in a continuity of past, present and future" (Adam 1990, 127). While past, present, and future are here conceived as continuous, a particular capacity of the human is noted as being able to transcend the present through retrospective and predictive capacities. Indeed, Adam involves the time-theory of Helga Nowotny in supporting the claim that no human societies "lack the ability to transcend the immediate present" (Nowotny 1975, 328).

By including all human societies in this kind of characterization, this perspective could be used to critique interpretations that equate certain cultural times with exclusively present-centric conditions. Rather than adhering to the numerous conceptions in which African time, Caribbean time, or Filipino time are restricted to a present-centrism and an associated ignorance in terms of lateness protocols, Adam and Nowotny appear to suggest that any culture has the capacity to collectively conceptualize beyond the current event.

In another regard though, I am concerned about how the phrase "to transcend" could inadvertently distinguish between an active internality, and a passive externality, to temporality. While Adam's and Nowotny's respective uses of the term "transcend" might appear to connote an active state or a dynamism that breaks from the jurisdiction of the exclusively present, what I actually interpret is a consequent passivity in no longer being embedded within the co-constitutive continuities of time. Adam and Nowotny have the intention that to transcend the present is to be able to actively and developmentally conceive of a past or future beyond the present. However, I contrarily identify a docility of the (hypothetically) transcended state, in that if time-consciousness has transcended the present, it can now only be *constituted by* the state (either past or future) to which it has transcended. In transcending the present, such a state cannot actively contribute its present situatedness to a *co-constitution with* past or future.

I will incorporate Habermas's characterization of critical versus passive states of time to further illuminate my point. While, as discussed in the first chapter, we have seen theorists within the Frankfurt School critique the rationalizing and bureaucratizing elements of capitalist structures, in this chapter we have encountered Habermas's seemingly *correlate* capitalist and critical imperatives. It is via this latter point that we can expand on my suggestion that Adam's and Nowotny's suggestions of transcendence separate internally active and transcendentally passive temporalities.

For Habermas, while the earlier reviewed arenas and discourses of the eighteenth century shift a representationally reactive or passive audience into a critically proactive one, the consequently highly engaged public sphere eventually deteriorates as a result of the influence of particular aspects of capitalism (including a growing commercial mass media). The problem that Habermas recognizes is that capitalist mass media, in *transcending* an era of the representational presence of power, actually returns the public (now taking the form of a population of consumers) to a state of relative passivity (Habermas 1991, 159–75). As Habermas states in the chapter "From a Culture-Debating (kulturräsonierend) to a Culture-Consuming Public," the critical mode that had been initially apparent in the eighteenth-century public sphere did *not* feature a transcendence of representational presence but rather an ongoing critique of, and engagement with, it. This sense of the interplay between the representational era and the critical era is replaced, however, with a new era of presence—capitalist presence. This mode supersedes and escapes, rather than engages and reconfigures, the previous representational form, whereby "the critical discussion of a reading public tends to give way to 'exchanges about taste and preferences' between consumers—even the talk about what is consumed, 'the examination of tastes,' becomes a part of consumption itself" (Habermas 1991, 171).[4]

While Adam's later text *Timewatch* acknowledges that we can "reinterpret, represent, restructure and modify the past" (Adam 1995, 38), her persistent characterization of this process as the human capacity to "transcend" the present differentiates her concerns from the sense of criticality that I believe is required in engaging the social constructionist separatisms of time states. I endorse Adam's protestation against "the tradition of understanding the 'then and there' as other to the 'here and now'" (1995, 169). However, by describing this as a present that is "transcended," such a conceptual division appears to be reinstalled, or repositioned, rather than reconfigured. Mead is occasionally also inconsistent in this regard, describing a temporalizing process "which transcends the present" (Mead 2002, 54). This is not necessarily a denunciation of Adam or of Mead but a reminder of the specificity of my focus. In the terminology exhibited by Adam, as well as briefly by Mead, suggesting that certain time states can transcend or be transcended by other states, the social constructionist condition lurks that elsewhere keeps prior (naturally real) and present (socially constructed) time states at a mutual distance.

I have not, in any case, correlated Mead's philosophy with transcendence. Mead has facilitated the insight that an individual belonging to more than one system (present-as-past-expression and past-as-present-production) does not transcend one system to be in the other simultaneously. Rather, one is only in a system because of their relational particularity to/in/as the other system. This (re-)produces both systems, present, past, immanently, concurrently and co-constitutively. In fact, the individual for Mead must be "contemporaneously in different systems to be what it is in either" (Mead 2002, 86). This "socialization," blurring apparently finite subjective boundaries and problematizing the notion of an eternally permanent finality of a subject's past, is conditional upon the present for its emergence (2002, 86). A reflection upon the capacity of Mead's philosophy to reconfigure the present-centrism found in a temporal model such as that offered earlier from Augustine, must therefore be qualified by the reminder that Mead also posits a present-centric rhythm, the "sociality of the present" (86). For Augustine, as we have seen, it is in the present that time upsurges. However, Augustine's motivation for interpreting time this way can be differentiated from Mead's when we observe that rather than defining the present according to a systemically inter-temporal sociality, in determining how the past and the future are dependent upon the present, Augustine turns to one's mind.

When considering what is past, Augustine notes that a "past time . . . no longer exists. But when I remember . . . it is in the present that I picture it" (Augustine 1961, 267). By engaging Mead we have challenged the supposed non-existence of *any* past, whereby there is never a past in-itself, but always a relationally produced "present-past." However, we could actually note that even for Augustine, in terms of what appear to be subjective past-present relations, while the past has passed, it is still reanimated via memory. The past thus has a conditional, ongoing, present existence as traces of experiences that "left an impression on our minds" (1961, 267). In remembering them, these pasts continue to be present. Furthermore, regarding the foreseeing of what is yet to occur, Augustine posits that "it is only possible to see something which exists" (268). This means that it is not that one predicts something that is yet to come into being, but rather that what is anticipated is already in the process of being (268). The earlier suggestion of the nonexistence of certain time states seems to be contradicted here, *if* thought anticipates a future that is already tracing and traceable in present indications. This is evocative of Pierre Bourdieu's earlier characterization of an African time that does not contain an abstract future. While it was noted that various futures or outcomes are possible from any particular present state for the Algerian culture he studies, any such future is always a development contained in the present, and as such there is nothing abstract or unexpected about it (Bourdieu 1963, 62).

Emphasizing the mental, or thought-based, present-centric constitution of the flow of time, Augustine clarifies that "the present of past things is the memory; the present of present things is direct perception; and the present of future things is expectation" (Augustine 1961, 269). The mind is also attributed with responsibility for how the future or past can be diminished or increased when they do not exist in themselves. Building on the previous paragraph, this point further coheres with Mead's refutation of a past in-itself. A differentiation of Augustine's position though seems to be his subjective-centric view of time's inter-state relations, affirming that "it can only be that the mind, which regulates this process" (1961, 277).

In raising this mental-centrism of present-centrism, it is worth asking how Augustine's prioritization of subjective, mental processes, can be accommodated within the structuralist argument developed in the earlier discussion in which individual minds are not prioritized or defined as demarcated, causal nodes, representing exclusively subjective modes of production. In response, we can discuss the notion of structuralist origination. For Durkheim, social currents manifest in/as the collective, and do not "originate in any one of the particular individual consciousness" (Durkheim 1938, 4). However, even if the subject is not an autonomous instigator of their behaviors, all individuals' behaviors collectively produce social rhythms. Given the lack of individual pre-enactment, the socially situated subject only becomes, and behaves as, that subject, in the present along with the social. Durkheim correlates such social rhythms with consciousness, as a "*state of mind* transmitted to . . . [a] number of persons [who] make the *state of mind* become an act" (Durkheim 1952, 273; my emphasis). In order to frame a return to Augustine, this discussion regarding the temporality of states of mind/consciousness brings us to a second feature of Durkheim's thesis that requires immediate investigation, complementing earlier concerns about possibly sequential timings between the socially objective and the subject. Now we specifically target whether the social fact of a human rhythm is objective because the collective consciousness from which it derives is outside individual consciousness.

Durkheim characterizes consciousness as manifesting socially "beyond" the minds of individuals. This is a *collaborative* consciousness that regulates what becomes factually continuous about human rhythms, whereby "the consciousness of individuals, instead of remaining isolated, becomes grouped" (Durkheim 1952, 275). Durkheim would therefore disagree with Augustine if the latter asserted a separate, atomic, subjective conception of time by the mind. Inversely, Durkheim would be more receptive if such subjectivity/mentality was portrayed collectively. This captures the porosities of Durkheim's model, of an external force that can only exist within the internal expression-of-the-collective that is the individual. While the behaviors that produce the uniformity of social rhythms are not individually isolated, they do manifest individually, Nick Crossley recognizing that "even if social structures are 'external' to individuals in the specific manner outlined by Durkheim, we should recognise, as indeed Durkheim did, that they never exist anywhere but within matrices of concrete human action and interaction" (Crossley 2001, 321).

By Durkheim describing the rhythmic uniformity of behavioral patterns as "social things, products of collective thought" (Durkheim 1995, 9), and recognizing that an individual constitutes, and is constituted by, such collective thoughts, a decentralization of the individual consciousness occurs. The products of collective consciousness, of which time is one, manifest regularly and even uniformly, coercing the knowledge of, and production of, social time, for all subjects "of the same civilization" (1995, 10). That different "civilizations" have different social times is, of course, a hallmark of social time for the social constructionist perspective. While, therefore, social time is "objective" for Durkheim, and does not misrepresent an "out-there real time" as the social constructionist time condition requires, its differentiated "objectivities" between concurrent global civilizations, or within the same civilization over time, exhibit an important constructionist perspective.

What I believe my reading of Durkheim's "version" of constructionism can offer is a simultaneity between individual and social consciousness. This reconfigures the social-individual sequence, by arguing that individuals do not straightforwardly arrive into an irrevocably originated, transcendent, and real apparatus of temporality but rather are participants in its construction. All individuals ever know is this socially manufactured time, because that is all that time is. Such an insight regarding the simultaneity of individual and social causation seemingly facilitates a more nuanced appreciation of the "socially constructed" body. According to my sense of Durkheimian structuralism, the decision to body modify does not objectively precede social adjudication, nor does social influence exclusively and transcendently precede individual decisions. Rather, there is a simultaneous co-production of society and individual, where subjective, individual decisions are implicated in/as social forces. These social forces are concurrently "outside" the individual, as objective impressions upon all in society, and "inside" the individual, via the incarnation of particular, objective expressions. The subject is outside its individual self by being inside its collective self, as a constitutive social element that is always already producing its individual self. Consequently, the problematic manifesto of the body modification community, which instills notions of a demarcated control over corporeality, becomes even more fragile. In the context of the compulsion to tattoo, to commit suicide, or to do anything, according to this reading neither society, nor the subject, can be said to exclusively predate, or arrive secondarily to, the other.

If we state that an individual is socially impelled to act with, and as, a particular social structure's present "objectivity," do we return to considerations of the negation of their agency or subjectivity? In response to this question, it could be posited that if subjectivity was absent, there would not simply be regularity or uniformity to social rhythm/temporality. Rather, there would be a straightforward *replication* of one social period in the next, producing identically molded "subjects" and a truly pure presence. That this does not occur correlates with my belief that there is still something uniquely localized about each individual's subjective time. This claim, of the social "individuation" of time, actually bears one of the hallmarks of social constructionism, in appreciating that no version of social production is the only, or the inevitable, version. Time could be constructed in many different ways, and indeed is, by different individuals. How, in the coming chapters, philosophies of time handle individually distinguishable times, will be crucial in appraising their critical efficacies regarding the tense distinction in the constructionist time condition between subjectively lived times and a supposedly separate reality of time. As noted, this should engender an investment in this inquiry from all readers, given that each of our times has an experiential reality beyond that of a mere representation of what is actual about time.

Contrary to Durkheim's definition of time's social reality, Augustine in fact disagrees that time is ever "objective," declaring that "I must not allow my mind to insist that time is something objective" (Augustine 1961, 276). This is seemingly an issue, given that "something objective" about time is required for social synchronization. How we know, or remember, to be "on time" for example, requires us to synchronize with a collectively standardized rhythm rather than to remain exclusively within the domain of a subjectively particularized realm. In addressing the objectivity of collectively conceived time and our subjective relation to it, we can consider conceptions of time-as-mind that might exhibit coherences with the position of Durkheim, as well as with the position that is albeit alternatively derived by the social constructionist time condition, that the time we live is always a social production.

COLLECTIVE TIME CONSTRUCTION

In terms of a subject's mental composition of time, one's straightforward past would suggest that memory is history is time. However, Augustine also offers a conception of a Durkheimian-like collective consciousness to illustrate that one's memory is not a mere "storehouse" of past, historical experiences. Rather, memory extends beyond a subject, who acquires and incarnates memories through present experience. Augustine acknowledges the social fabric of time-consciousness in doubting his ability to "think of them [memories] at all, if some other person had not brought them to the fore" (Augustine 1961, 218). Rather than the subject or the past memory preceding the other, here subjectivity seems to be a socially dispersed manifestation of time-memory-consciousness. What occurs is the becoming of the subject, where memory, in *collectively* bringing forth that of which one is not always aware, manifests as the process "where I meet myself" (1961, 215).

I will suggest that Augustine here evokes Mead's co-constitutive past-present relation, in that the subject who remembers the past only manifests in the present concurrently with the past that also manifests. Mead characterizes the mind in equally broad, collective terms, stating that "the field of mind is the larger environment which the activity of the organism calls" (Mead 2002, 54). Given the materially/physically organismic element of Mead's argument, in the philosophical critiques to come in this book of perspectives that divorce materially natural time from socially representational time, his philosophy will again become relevant. For now however, in continuing to identify foundations of the social constructionist tendency in question, if we are to benefit from engaging a model of subjective time, a socially unrestricted conception of subjectivity is necessary in order to explain the common/objective time-consciousness that conditions experiences of an individual's social synchronization and lateness. This was noted at the end of the preceding section. Interestingly for Augustine, the mental constitution of time-consciousness is a socially conceived phenomenon. This does not suggest that the time we think and experience is a fabricated misrepresentation of a real, "out there" time. Nor does it separate a socialized source of time from the individual compositions of it, as was our concern at the start of the chapter regarding Durkheim. Rather, it posits that our only experience of time-consciousness requires an interpersonal participation.

This recognition of subject-social porosity is congruent with the work of one of Durkheim's peers, Maurice Halbwachs. In *On Collective Memory* (1992 [1941]), Halbwachs's argument is that "collective memory is not a division, or a given, but rather a socially constructed notion" (Halbwachs 1992, 22). Individuals acquire *and* produce their memories socially (1992, 38), whereby memories reflect our relations with others, just as Augustine argues that memories would not even manifest were it not for other people bringing them "to the fore." In belonging to such memorial contexts, Halbwachs notes that "when I remember, it is others who spur me on, their memory comes to the aid of mine and mine relies on theirs" (38). Recollective processes are attributable to one's presence in a social frame. In fact, it now becomes apparent; to re*collect* is a *collect*ive operation, or, dare I say it, a recollective *construction.*[5]

In terms of our current exploration into the relation between socially constructed and individually constructed times, I argue that another way to describe a collectively constructed consciousness is that the subject (re-)produces the social that (re-)produces the subject's memories. This (re-)emergence of collectively conscious states, in the form of states of temporality, is apparent in Halbwachs's identification that "collective historical memory has both cumulative and presentist aspects," showing "partial continuity as well as new readings of the past in terms of the present" (Halbwachs 1992, 26). Collective memory embodies social structures that subjects perpetually inhabit, whereby something like a suicide could be said to participate in/as the memory of a social body via its contribution to the overall, uniform, rhythm of suicide rates. The social body *remembers itself* in a blurred movement from the specificity of the individual to the collectivity of the social, incarnating individuals who incarnate its, and their, rhythmic production. This continuous uniformity "reminds" subjects in subsequent social periods how and when to take their own lives. The Durkheimian "mystery" of social uniformity might here be articulated. If each subject does indeed play the role of a continually socially memorial constituent, I would argue that the interpretation of suicide (or of death generally) as marking the finality of one's social participation is debatable.

Where this situates the subjective consciousness of time, and of lateness protocols, is consistent with earlier theories of the social production of subjectivity. The collective consciousness of what it is to be social will differ depending on the cultural context. We have observed this in anthropological accounts of different regional lateness protocols. Nevertheless, within each context, a Durkheimian reading could be that the inclination for an individual to be late, or to have a general disregard about being on time, does not represent an external contestation to what is socially structured and normalized about time. Rather, the source for lateness must materialize from the same source as promptness and represents the myriad ways in which the social organism manifests its shifting consciousness of temporal politics. That these myriad forms are, for the specificities of the social constructionist time condition, an indication of the constructed separation of the time that the subject experiences, from an inaccessibly real time, can now be made further apparent via relatively recent portrayals of subjective time.

This returns us to the work of Michael Flaherty. We can identify that a conditional navigation toward Durkheimian structuralism is apparent in Flaherty's definition of subjectivity in social terms, stating that the "self is a thoroughly social entity" (Flaherty 2011, 132). However, Flaherty, counter-Durkheimian, relies upon a resiliently interiorized characterization of time-experience, in presenting a subjective mind's "internal environment (i.e., self-consciousness, cognition)," which interprets "information from its external environment (i.e., one's situation or circumstances)" (Flaherty 1999, 141). Complementing this juxtaposition of the mind's internality from a social, situational realm is his impression of one's mind as a separate container that can gain and lose memories. In cases of "temporal compression," which is Flaherty's term for experiences in which there is "less" conscious experience in a "standard" period of time than is typical (1999, 105), there is hence said to be a "loss of memory over time" (110).

Flaherty's quantification of memories invokes the conception of a storehouse that Durkheim and Augustine avoid. By focusing on there being more or less isolated memories, the past is separated from a present in which the past, as an objectively fixed fact, is either retained or involuntarily relinquished (Flaherty 1999, 110). This contradicts my Meadian inspired reading of the ever-relational past and can be categorized with the social constructionist position under examination in which a real past-presence of time is contingently misrepresented in subsequent presents. For Flaherty, time is quantifiably objective because its source is outside each present subject, a preexisting, historically originated utility (1999, 35). The natural-then-social-then-individual sequence is perpetuated in this model.

Eviatar Zerubavel's conception of collective memory is seemingly more consistent with a critique of the constructionist tendency to separate the individual from what time actually or really is. Zerubavel's *Time Maps: Collective Memory and the Social Shape of the Past* (2003) evidences this, exploring a "pronouncedly social dimension of human memory, by revealing how entire communities, and not just individuals, remember the past" (2003, 2). In acknowledging that we remember collectively, Zerubavel mimics Halbwachs's and Augustine's description of the socially environmental drivers and tempos of individual memory (2003, 4). It is following this, however, when arriving at the originality of Zerubavel's work, that a tension manifests.

Congruent with the argument developed in this chapter with Mead, Zerubavel acknowledges that the past and the present are not absolutely distinct, nor "entirely separate entities" (Zerubavel 2003, 37). Zerubavel attributes this to the present amassing what has preceded, whereby "the present is largely a *cumulative*, multilayered collage of past residues" (2003, 37; author's original emphasis). However, Zerubavel also characterizes impressions of past-present co-implication/continuation as an *illusory* mental technique that combines or "bridges" what are actually non-continuous points in terms of anything's changing states over time. This argument is that in constructing a sense of temporal continuity, we actually know that we are ignoring "temporal gaps" (40). Such a claim is difficult to reconcile with his previous reading of a past that is not separate from the present, instead now stating that "continuous identities are products of the *mental* integration of otherwise disconnected points in time into a seemingly single historical whole" (40; author's original emphasis).

The reason for engaging Flaherty, and now Zerubavel, is not to outright criticize either. Rather, this discussion complements the ongoing first stage of this book, designed to identify the social constructionist time condition's separatist orientations. Zerubavel's argument that the mind illusorily arranges a version of time, assumes, I posit, that time's reality is inaccessible to a consciousness that subsequently generates representations of it for socially pragmatic purposes. This is consistent with the interpretation that real time is not the social construction that humans structurally and mentally experience but rather is a celestially occurring phenomenon, anterior to its human representation and utilization. In this style, according to Zerubavel, the individual employs techniques to construct a useful impression of time via "mental editing" (Zerubavel 2003, 40).

Flaherty's notion of temporal compression, and Zerubavel's sense of time's mental construction, bring empirically validated accounts of the diversity of subjective time-experiences. Rather than undertaking my own empirical studies in this book, such conclusions are integrated because they exhibit the belief in a pluralized subjective construction of time that often occurs in tandem with contingent social structurations. Our primary investigation is served if their inclusion further illustrates distinctions that are made between these representational times and a real time on which such representations are based and from which they diverge.

In such modes we see that the time of the subject can only ever be a socially coerced abstraction. Of course even in Durkheim we are given the impression that time is only a social idea. The difference between Durkheim's work, and that of a theory such as Zerubavel's, is that the latter focuses on the human experience, representation, and knowledge of contingent symbols of real time. Conversely for Durkheim, such symbols, and one's consciousness of these symbols, while being entirely social, do not *misrepresent* real time. In Durkheim's structuralist perspective, social time is really time. This conclusion, as has been noted, presents a certain incoherence with the strict sense of the social constructionist time condition's separation between naturally real, and socially contingent, times. Despite this qualification, Durkheim's symmetry with the constructionist condition is still evident as previously explained, in that our time for Durkheim can only ever be social, even when being lived as an "objective thing" in any one of numerous social contexts/civilizations. Furthermore, the concern has been raised that for Durkheim, as for the social constructionist position, the real source of time is externalized from individual consciousnesses of time.

In situating Durkheim among recent commentaries on this point, one encounters a plethora of characterizations of Durkheim's constructionist tendencies. For Randall Collins, Durkheim is "the great social constructivist," given that it is only via social events that categories of human experience manifest, around which "humans have constructed a symbolic scheme of time" (Collins 2007, 26). Similar impressions are found in Misztal (2003), Loseke (2017), and Fish (2017), the latter noting Durkheim's tangential characterization as an architect of the social constructionist approach to the study of emotions. Michael Halewood provides an interesting twist on such impressions, by suggesting that the "fact of construction is not a problem for Durkheim, as long as it is realized that construction is not limited to the social realm. Construction itself is natural" (Halewood 2014, 53). While

Halewood's point *might* be valid that nature is able to construct itself for Durkheim, the social construction of reality for Durkheim is not straightforwardly nature's doing. Such social construction occurs in the form of categories in the Durkheimian perspective, of which time is one, that fall somewhere between the empiricist impression that "categories are purely artificial constructs" of nature, and the *a priorist* impression that "they are works of art . . . that imitate nature ever more perfectly" (Durkheim 1995, 17). While Durkheim is not the kind of constructionist to state that social categories simply serve us "in conceiving the physical or biological world" (1995, 17), social time in his thesis is qualitatively differentiated from a truly natural counterpart. Just as we have seen the social constructionist time condition characterize nature's time as straightforwardly singular versus social times' complicated pluralism, a social category such as time for Durkheim represents "nature's highest expression" that differs according to "its greater complexity" (17). Durkheim indeed concedes that if he was to accept the empiricist's characterization of the "artifice" of the social category of time, this must only be as a construct that "closely follows nature and strives to come ever closer to nature" (17). For the reasons discussed in this chapter, I do not position Durkheim as the kind of social constructionist who separates social time as an approximation of real, natural time. However, there is enough in his characterization of time that nonetheless coheres with key parameters of the social constructionist time condition.

We have, in any case, been concerned with the social-subject sequence in this chapter. In terms of the positions established during the preceding Durkheimian and Meadian discussion, the development of an argument that situates individual consciousness/subjectivity in the temporality of an objective, collective rhythm, is important when engaging the social constructionist time condition. This is because, beyond Durkheim, it facilitates an interrogation of the general assumption of a sequence in which firstly there is a naturally real time, *then* there are separately contingent, socially constructed, times, *then* there are externally myriad, individual experiences, of this socially constructed time. As we have seen, one ramification of this assumed sequence is that rather than the "cultural relativization" of socialized times fulfilling a critical, liberationist mandate, by recognizing the respective politics of every culture's time structure, instead those living within certain structures are restricted to characterizations more closely aligned with a pre-social, natural state.

While the sequence between the individual and the social has potentially been opened, Durkheim's characterization of the time we experience, as always bound to social phenomena, does not shift the natural-social polarity of the social constructionist time condition. Durkheim does not necessarily perpetuate the social constructionist time condition's separation of naturally material/physical time and socially representational times, but rather just ignores it. There is no negation in Durkheim's thesis of the role of materiality/physicality in the temporality of socially uniform rhythms (suicide is, after all, an embodied phenomenon). Indeed, at one point he explicitly refers to how individual manifestations of collective conditions "depend to a large extent, on the *organopsychological* constitution of the individual" (Durkheim 1938, 8; my emphasis). Such organic dependence is discussed in Chris Shilling's examination of Durkheim's study (Durkheim 1995) of totemism. Shilling astutely argues that the individual body, rather than being ignored by such inquiries, is instead "inescapably anchored in the natural world as a critical medium through which the symbolic order of society, or the social body, is constructed" (Shilling 2005, 213). Furthermore, consistent with Shilling's observation, is Durkheim's own, albeit somewhat isolated, claim in his essay "Individual and Collective Representations" (1974 [1898]), that there is no need to separate an ideal milieu from the body (Durkheim 1974, 28).[6] Nevertheless, the literal physicality of temporality is nowhere thoroughly explored by Durkheim, instead relying upon the definition of time as a socially conceived category.

This marks the culmination of the stage of this project that is concerned with the *identification* of the social constructionist time condition. While we have interrogated the conceptual, sequential separation of social objectivity from subjectivity, the condition stands that time is either singularly natural *or* plurally social. Our attention is now to be directed to whether philosophies of time are able to destabilize or disrupt the social constructionist time condition's separatisms. In this chapter, the individual has been situated as a necessary constituent of social time. Potentially, therefore, the primary investigation into what might be common, rather than opposed, between naturally material temporality, and socially structured temporality, can first attend to what is naturally material about that socialized/socializing individual—their body.

Chapter Four

(De)Constructed Bodies

Are Modifications Late to the Corporeal Scene?

Human beings make meaningful the world which makes them.
—Loïc Wacquant, "Toward a Social Praxeology" (1992, 7)

STRUCTURALISMS

It is at this point that we reach a methodological juncture. Engagements with various sociological, anthropological, and even philosophical, perspectives on time have identified the social constructionist time condition that separates natural time and social time. This has been largely made distinguishable via an analysis of the cultural relativization of lateness regulations, in comparison to the conceptualized singularity of natural time. A second context, the interpretation of the timing of socially modified and constructed bodies, in relation to the natural body from which they derive, has also been opened. Lateness regulations, and body modifications, are two of the four contexts through which is enacted the primary investigation of the book into the social constructionist separation of natural and social temporalities. For readers seeking a reminder of this book's structure, these distinguishable, but complementary, contexts are outlined at the end of the introductory chapter.

It is in responding to the prevalence of the social constructionist time condition that from this point we will be considering whether the "philosophies of time" also noted at the outset of this undertaking not only identify the polarization of natural and social temporalities presupposed in their theoretical preconditions, but intervene into and dismantle it. While accounts from philosophy *and* social science have been integrated in the first three chapters, the exclusively philosophical orientation of this critical reflection becomes apparent hereon. This occurs in this chapter via a continued, but more interrogatory, focus on the context of the body. Subsequent chapters will then explore the two remaining contexts, concerning the perceived lack of timeliness in our response to climate change, and the possible secondariness of theoretical representation in relation to the material reality it represents.

The capacity of philosophy to reconfigure divisions of natural and social temporalities is in this chapter explored through the deconstructive "method" of Jacques Derrida. Deconstruction can be characterized as a post-structuralist critique of structuralisms, offering the notable quality that instead of presenting its own system of thought, deconstruction reveals the contradictions inherent to the theory being engaged. As we will encounter, such contradictions are typically presented in sets of binary pairs that Derrida identifies within the structures being analyzed, critiquing the priority that is given to one pole of any such pairing. It is via this method that, in the context of the body, we are able to attend to the polarized pairings of body-as-material|mind-as-not-material, and natural-material-time|social-not-material-time.

The conceptual separation of body from mind is particularly relevant, given our preceding review of the characterization of the body modifier as a cognitive agent who possesses power *over* their corporeality. It is from such readings that socially informed, subjectively enacted, body modification processes are seen as arriving late to one's natural corporeality. This temporal sequence between natural corporeality, and subsequent or late interference, encapsulates the secondariness under examination in this book. From this we should be able to appreciate why this context of the body helps us to unfold the impression that socially structural representations of reality are divorced from the material reality upon which they are based.

In order to frame the Derridian intervention according to its critical endeavor, we need to consider the theoretical intersections of structuralism and criticism that precede deconstruction. What is first relevant here is the field of "New Criticism." Manifesting as a response to the dominance of historicist, literary approaches, which were perceived to be distracted by the biographical and sociological parameters of a text being engaged, New Criticism refocuses literary studies on direct textual analysis. Leroy Searle accordingly describes the motivation of New Criticism's key proponents to "focus critical attention on the literature itself" (Searle 2005, 691).

While there is no formal association of "New Critics," commentaries generally agree that John Crowe Ransom, Allen Tate, Cleanth Brooks, and Robert Warren develop the style that would eventually be known as New Criticism.[1] This New Critical attention is navigated in such a manner that a structure of meaning must be considered to be internal to the text in question. New Critics advise that the study of literature should no longer accommodate subjective taste or environmental considerations but instead be undertaken with the rigor of a professional academic discipline, whereby "its proper seat is in the universities" (Ransom 1937, 587).

What is seemingly most at stake for the newly critical model in evaluating a text is the exclusion of readers' opinions, speculations about the author's intentions, as well as considerations of the cultural contexts in which the text was written. Instead, by looking for an internal system of logic to literature, Searle says that New Criticism can be correlated with the defining parameters of structuralism (Searle 2005, 691). Furthermore for Art Berman, given New Criticism's structuralist characteristics, structuralism inversely can be considered a form of critical theory:

> In its application to literary criticism, structuralism appeared, at first, to meet the criteria demanded of a critical theory in the United States in the late 1960s: it was, presumably, an empiricist science; it incorporated linguistics, it addressed the problem of human mentality; and it issued from the political left. (Berman 1988, 171)

This scientific imperative of New Criticism is overtly evident in Ransom's foundational essay, "Criticism, Inc." (1937). Emphasizing the structuralist aspects to what he perceived to be the central mechanisms of critique, Ransom demands that criticism "must become more scientific, or precise and systematic" (Ransom 1937, 587). Here we can interpret that a prioritization of the systematization and structuration of this critique distinguishes it from the critical theory of the Frankfurt School. The position I wish to present on this theme is that while theorists such as Horkheimer and Adorno have been shown to be concerned, in a text like *Dialectic of Enlightenment* (Horkheimer and Adorno 2002), with how critical theory can intervene to rationalist systematizations of life, for a New Critic such as Ransom, systematization and critique are actually literary complements. This relates to the point that we have reviewed, that for the Frankfurt School there is an emphasis on historicizing the ideas of any particular field. This occurs by contextualizing the development of such ideas in order to dismantle any timelessly normalized power they might present, contrasting traditional theory from critical theory in that only the latter is interpreted to be receptive to its own developmental genesis.[2] Conversely for the New Critics, an idea in literature must be analyzed *outside* the distractions of historical and contextual parameters.[3] Meaning for the text is discoverable as that which is entirely internal to its presented structure.

We have encountered a form of structuralism in Durkheim's social facts. Durkheim's project is concerned with taking subjectivity beyond its demarcated, individual limits and into the social, structural substratum. In doing so, historical foci for the signification of individuals and social structures have emerged. I argue, however, that the historical element of Durkheimian, sociological structuralism, does not contradict the interpretation of an ahistorical character of linguistic structuralism. Just as the "text" is interpreted by New Criticism to have an internal logic that can be deciphered via the relations found within the developments of its single structure, similarly I posit that for Durkheim the social structure manifests with meaning only as one tracks an internal logic that radiates within it. The internality of the system or structure in either context can be here said to write its own "history." This history, however, is an enclosed, self-referring mechanism. It will soon be considered that by deconstructing this enclosed feature of structuralism, and of New Criticism, Derrida opens our considerations to the possibility of structuralist perpetuations of the social constructionist time condition.

Durkheim's structuralism in this regard can actually be viewed as correlating with the structuralist methodology that is embedded within important linguistic analysis. Here I refer to the semiological theory of Swiss linguist Ferdinand de Saussure. Presented as a science of linguistic signs,[4] Saussure's semiology distinguishes between *langue* (language as what people have in common) and *parole* (the speech of each particular person). In identifying an impersonal system of grammatical patterns and syntactical rules in language (langue), Saussure argues that it will then be possible to distinguish "what is social from what is individual, and what is essential from what is ancillary or accidental" (Saussure 1966 [1916], 14). By determining what is essential (social) within a system, the way common meaning is produced will be identifiable.

To explain this production, Saussure defines language as a *system of signs*. Each sign consists of a *signifier*, its descriptive tool (such as a word), and a *signified*, the related mental concept. Together, these components *contingently* comprise a sign. This composition is interpreted as contingent, or "arbitrary," because any number of signifiers/words could seemingly have been associated with any concept, in that "the signifier has no natural connection with the signified" (Saussure 1972, 73). That nothing about the construction of a sign is self-evident demands that its meaning and signification is only found by comparison with other signs within an entire language system. Meaning does not eventuate from a necessary correspondence between signifier and signified, or interestingly for our social constructionist considerations, between representational sign and the real world. Rather, it is a relational and differential mechanism between signs, just as for Durkheim, one's subjectivity and identity radiate through socially structural, rather than individualized, conditions.

It is on the basis of a continuing debate regarding the comparability of these structural parameters that Durkheimian and Saussurean intersections present as topical matters in recent scholarship. Patrick Baert and Filipe Carreira da Silva liken a system's *externality* from any individual node in Durkheim's sociology to the structure-node relations of Saussure's linguistics (Baert and Carreria da Silva 2010, 21). This is a theme that is reiterated by Hans Herbert Kogler (2011, 277). Perhaps more adventurously, Lynn Badia argues that while Saussure is often cited in standard, current accounts of post-structuralism, Durkheim is "entirely absent." This is said to be despite the fact that Saussure, along with post-structuralists like "Derrida, are indebted to Durkheim" (Badia 2016, 972).

My interest in Saussurean structuralism, however, expands in a different direction to any specific, theoretical intersections that might be observable with Durkheim. In fact, I do not want us to become too preoccupied with Saussurean theory. This is because, rather than engaging Saussure at length, his relevance emerges in appreciating how deconstructive criticism comes to develop a sense of what is at stake in structuralist logics of differentiation, in order to intervene where I believe the social constructionist time condition pervades.

One such deconstructive critique of Saussurean structuralism is identifiable in how the arbitrary relation between the signifier and the signified is reconfigured by the French linguist Émile Benveniste. Saussure argues that the contextual contingency of the signifier-signified relation is exemplified in translation, where different cultures' signifiers refer to the same conceptual/signified "thing" (Saussure 1966, 68). For instance, different cultures will have different words for the signified thing "cow." What this indicates for Benveniste, though, is that if many signifiers coexist with the same signified, then for Saussure the signified has an irreducible, preexisting quality, which each language expresses in its own way (Benveniste 1971 [1966]).

We might recognize qualities of the social constructionist time distinction between a singular reality, and plural representations in this analysis of Saussure by Benveniste. Saussure's "mistake," according to Benveniste, is seemingly attributable to his presumption that *access* to the material substance of reality is provided through signs that are outside such material, substantial reality. Given our interrogation of the constructionist position that inaccessibly externalizes the presumed secondariness of social *contingency* or *arbitrariness* from the material reality of time, Benveniste's attention to this structuralist condition is important. Indeed, while Saussure's semiology assumes that a world exists prior to language, Saussure holds that it did not exist meaningfully. For Saussure, before the differentiation/signification that language provides, the world is one anonymous plenitude, in that "nothing is distinct before the appearance of language" (Saussure 1966, 112). In Saussurean structuralism then, I argue that a perpetuation of the social constructionist assumption is apparent, in which a singular, natural, real phenomenon, is differentiated by plural, social, contingent forms.

When referring to the signified half of the sign, Benveniste believes that Saussure appears to have been concerned with its connection to the "real thing" in the world. This manifests in the evidently "unnecessary and unmotivated character of the bond which united the sign to the thing signified" (Benveniste 1971, 47). Here, we are told, is the real arbitrariness within Saussure's semiology, that already lurks unidentified. That the thing emerges only through the self-referentiality of the system indicates that what Saussure instead intended to argue according to Benveniste is that the relationship between the sign and reality, between the sign and the "real world" thing, is what is arbitrary (1971, 46). Armed with Benveniste's deconstructive critique, we can say that Saussure's semiology exhibits the issue encountered in previous chapters concerning social constructionist interpretations of time by positing an inaccessible reality that is contingently (arbitrarily) represented by human culture.

I would like to expand on this thought by asserting that Saussure's semiological structuralism, along the lines of Benveniste's deconstruction of it, provides a certain insight into why the social constructionist position might view social time as contingent. Social time symbols, or signifiers, such as those on clocks, could have been developed in any number of ways. Their connection to the "real thing" of time is interpreted therefore to be entirely arbitrary. Such symbols only come to have meaning via their relation to other, equally as contingent, time signifiers/parameters that never actually access the element of reality to which they refer.

There are readily found examples in recent debates about Saussurean semiology that posit that it excludes the "external, real world." This argument is typically that Saussure does not adequately convey how the sign relates to reality (Smith 2010), or that the Saussurean sign involves nothing akin to a referent or an object and gives such entities no role in the signification process (Nellhaus 1998). While I am receptive to these arguments, I am not necessarily positing in my interpretation that Saussure's semiology is definitively social constructionist in its overall composition. Indeed to make such a characterization would require much more engagement with Saussure's overall structuralist mandate than is necessary for my integration of the previous, key components. Rather, what has intrigued me in these discussions are the aspects of Saussure's semiology that exhibit tendencies consistent with the specific social constructionist time condition we are examining. Dave Elder-Vass is in relative agreement with this reading of Saussure, arguing in *The Reality of Social Construction* (2012) that while Saussure's linguistic constructionism is entirely social in constitution, we should "not license" all "constructionist conclusions" about it (Elder-Vass 2012, 79). This is a perspective with which Maze (2001), and Norris (1997), would seemingly agree, given their arguments *against* the post-structuralist, constructionist characterization of Saussure. For Elder-Vass though, and consistent with his general thesis that social constructionism and realism should not necessarily be conceptually polarized, what is especially important to recognize is that Saussure's conception of the langue|parole distinction is both constructionist *and* realist. While langue is presented with constructionist elements, the Saussurean indication of the linear temporality of parole coheres, for Elder-Vass, with what "a realist would want from an account of parole" (Elder-Vass 2012, 87–89).

In accordance with the relational construction of meaning for words in langue, a relevance emerges from Saussurean semiology for the two contexts that have been raised in this project—the relativization of different social times, and the socially contingent meanings of body modifications. In fact, while Saussure is concerned with an enclosed linguistic structuralism, he envisages that its semiological approach speaks to other realms with which the human sciences are concerned:

> *A science that studies the life of signs within society is conceivable.* . . . I shall call it semiology. . . . Linguistics is only a part of the general science of semiology. (Saussure 1966, 16; author's original emphasis)

Recent debates in cultural studies have adopted this directive accordingly. Semiological analysis is applied to interpretations of the componentry of visual "texts" involved in media (Stokes 2013), photography (Wright 2016), shopping (Miller 1998), fashion (Barthes 1990), not to mention the identity parameters of race (Bell 2018), gender (Butler 1990; Lloyd 2013), and sexuality (Brooks 1997; Schulz 2011). My reading of time and body modification can be situated alongside these works by exploring in this chapter how we can read both phenomena semiologically. The focus is not merely to indicate the semiological, relational production of either. Rather, the critical component of this stage of the inquiry also requires an investigation of the possible destabilization of the constructionist opposition of natural and social temporalities that might be apparent in the structuralist presuppositions targeted by deconstructive interventions.

Just as meaning is produced differentially within a language structure, similarly I believe that we must appreciate that the internal mechanisms of social time also show structuralist production. By revisiting the Nuer, we can see how their time refers to social activity. Any such activity is associated with the ecological change that is limited by an annual cycle, and "cannot be used to differentiate longer periods than seasons" (Evans-Pritchard 1940, 94). My reading of this is that times are *internally* signified, according to seasonally differentiated, associated activities, as the routine tasks of the dry season differ from the varied activity of the rainy season (1940, 103). The socially structured constitution of time emerges; while Evans-Pritchard does not recognize its production as *semiological* in nature, he is in agreement that "all time is structural" (104). The structural arrangement of Nuer time is also evident in their "age-set system." Age-sets describe not how long ago an event occurred in linear isolation but rather its relations to other events (105).

Structural time cannot be discussed without also referring to Durkheim. Time is entirely socially constructed in his structuralism, because "we can conceive of time only if we differentiate between moments" (Durkheim 1995, 9). These moments are socially conceived, and any particular one has no meaning in itself. Just as in the previous chapter it was argued that there is no past "in-itself," equally for Durkheim time would not manifest for us if it was "not divided and differentiated" (1995, 10). While, as discussed, Durkheim does not necessarily posit an inaccessible reality that social reality (mis)represents, all that is ever experienced temporally is a structured construction and categorization. Even space is a structural phenomenon for Durkheim. Particular spaces arise via differentiation from the rest of space(s), arranging/emerging "differently . . . some on the right, others on the left, these above, those below . . . just as, to arrange states of consciousness temporally, it must be possible to locate them at definite dates" (10).

If we are to grant the structuralist premise that all production is relational, then we can indeed revisit the prohibition of tattooing by certain religions. I am inspired here by the insight that according to a structuralist ethos, the "normal body" must manifest as normal only in relation to other bodies and "not-bodies." This suggests that such proscription of certain forms of body modification is not an insight into the inherently evil character of a socially informed practice such as scarification, for example, but instead simply represents what is meaningful for one religion in comparison to another. While dermatological scholar Noah Scheinfeld does not recognize this as a semiology, he nonetheless interprets according to a structuralist mechanics the biblical citation from Leviticus upon which Christianity and Judaism have equated bodily marking with sin. For Scheinfeld, any opposition from one religion toward body modification is underpinned by how every religion distinguishes itself from other religions (Scheinfeld 2007, 363).

A structuralist perspective duly complements and develops the contestation to characterizations of non-normative modifying practices being either a defiantly individual control, or a subsequently late hijacking, of an already established corporeality. A structuralist appraisal would be that while body projects appear to be self- or secondary productions, intervening where nature has a prior integrity, they actually only manifest within the already existing organization of a social system. This resembles what anthropologists observe in many tribal cultures where tattooing speaks of collective rather than simply of personal expression. Alfred Gell notes that "in traditional Polynesian settings the tattoo was significant not so much as a thing in itself, than as the relative social standing it declared" (Gell 1993, 305–6). Correlatively, something like a piercing is linked to all other piercings, and non-piercings, and meaningfully constructed accordingly. Eric Gans observes that in any culture, in being pierced, individuals are "nodes whose privilege is organizational rather than hierarchical," whereby "the pierced body as a whole holds no particular interest in itself, it is merely a hub from which to radiate signification" (Gans 2000, 159). Gans indeed takes his analysis into specifically semiological territory, in that the modifications of the body are posited as only being able to be read according to the "semiotic character of their inscriptions . . . whereby the pierced or tattooed body exhibits the arbitrary meaningfulness of the inscribed sign" (2000, 159).

According to a structural perspective, therefore, time, material space, corporeality, indeed everything, can be interpreted as manifesting differentially within the unity of a singular structure. With this recognition of the differentiation that occurs within a structural unity, we return to New Criticism's systematic appreciation of the text as an internalized arena of meaning, a feature that Leroy Searle describes as "the formal unity or balance of the work" (Searle 2005, 5). New Criticism's, and indeed structuralism's, interpretation of a text's or a system's unified balance is however not adopted by Derridian deconstruction. Derrida's contrary focus is on destabilizing structural meaning. The coming Derridian debate should attract the interest of readers interested in exemplifications of philosophy's critical utility. This is due to how, in destabilizing structural meaning, we could also be unsettling the social constructionist time condition's essentialized differentiation between natural and social phenomena.

DERRIDIAN DIFFERENCE AND HIERARCHY

As Durkheim's structuralism enquires into the differential production of time and space, he asks, "[W]hat is the origin of that differentiation?" (Durkheim 1995, 9). Given that "origins" are a pertinent issue regarding the constructionist sequence between a transcendent real/natural time, and relationally representational social times, we will benefit from Derrida's development of structuralist concepts on this theme.

For Saussure we should recall, neither a signifier nor a signified has positive content. Instead, each is generated via its simultaneous difference from other signifiers and signifieds. Derrida's critique of this structuralist assertion concerns the status of *difference* itself, suggesting that to perennially condition the differentiation of *signifiers from signifieds*, difference is given its own origin. As a consequently "transcendental signified," this status of difference contradicts the argument of a simultaneously stable production of *all* signifieds:

> There has to be a transcendental signified for the difference between the signifier and the signified to be somewhere absolute and irreducible. (Derrida 1976 [1967], 20)

Aspects of recent literature debating Derrida's targeting of this transcendental status of difference accept the deconstructionist's critique, while arguing that this does not necessarily discount Saussure's perspective. As Russell Daylight posits in *What If Derrida Was Wrong about Saussure?* (2011), Derrida's argument might exceed Saussure's requirements and go beyond semiology's capacity to respond. Accordingly, Daylight posits that Derrida simply explores a complementary thesis on the systemic condition/constitution of difference, in that "when one takes the position that the constitution of a system must precede the system, then certain truths are possible" (Daylight 2011, 170). This does not contradict the converse, Saussurean interpretation, however, for "when one takes the position that the conditions of the system can be situated as objects within that system, then other truths are possible" (2011, 170).

This is possibly a reasonable critique, as Derrida's commentary on the transcendental signified attends to the structuralist's interpreted supposition, rather than actual integration, of it. Michael Peters and Gert Biesta additionally note that the deconstructive intervention could even bring a reflection on the impossibility of transcendental signification itself, if "it is acknowledged that there are no simple, unsignified, transcendental signifiers that fix and warrant the meaning of our words, there are no originals to which our words can refer" (Peters and Biesta 2009, 25). Every transcendental signified, to be articulated as a presence, or the origin of a system, would itself need to be signified. The signified must signify and perform the role of a signifier, simultaneously, a point that Derrida himself makes in recognizing the consubstantiality of the sign:

> From the moment that one questions the possibility of such a transcendental signified, and that one recognizes that every signified is also in the position of a signifier, the distinction between signified and signifier becomes problematical at its root. (Derrida 1981 [1972], 20)

If there is nothing positive about the identity or meaning of a signified, not even the condition of difference by which it refers to other signifieds and signifiers, then Derrida must explain the production of meaning in a different manner to that which he has critiqued in Saussure. Derrida accordingly introduces *différance* in *Margins of Philosophy* (1982 [1972]). Différance is a reference to the differential production and meaning of anything, as "a detour, a delay, a relay, a reserve, a representation—concepts that . . . could be inscribed in this chain: *temporization*" (Derrida 1982, 8; author's original emphasis). This is the noteworthy character of différance, deconstructing structuralist thought by not simply arguing that meaning is possible relationally but arguing that if meaning or identity is produced through relations alone, then it never emerges permanently, fully and finally (1982, 9).

The possibility of meaning ever emerging at all has permeated contemporary debates on Derridian deconstruction since. David White explores this issue in terms of the negation of the transcendental signified, wondering "If chains do function in this regard, how is the process accomplished?" (White 2017, 139). The point I argue for Derrida is that "accomplishment" or presence is *not* reached. White is, however, not convinced, arguing that because a "chain" establishes significance for any of its component parts by establishing an integration between such parts, it "seems necessary that a chain construed as some kind of organic unity must win access to something like a transcendental signified" (2017, 139). I am quite fond of the kind of response offered to this interpretation by Raymond Tallis, who notes that the fact that the signifier is "not accomplished" as a transcendental signified "should be a cause for concern only if the signifier were the sign itself and the 'transcendental signified' were a referent. And manifestly they are not" (Tallis 1997, 463). Instead, as Tallis asserts, Derrida uses the concept of the transcendental signified to move from the position that no sign opens directly to the "plenitude of meaning/presence" in the real world, to the claim that there is no timelessly separate signified to which the socially constructed signifier can reach.[5]

By maintaining the structural element to Saussurean semiological analysis, Derrida's différance fulfills the aforementioned requirement of deconstructive critique; to avoid negating what is fundamental about an established argument, but instead to unpack or navigate that argument toward its supposedly actual purpose. As Derrida notes, the deconstructive reading of différance is thus inherently linked to structuralist positions, whereby "to deconstruct is . . . a gesture that assumes a certain need for the structuralist problematic. But it is also an anti-structuralist gesture, and its fortune rests in part on this ambiguity" (Derrida 1985, 2). It is in such a regard that deconstruction can be considered to exhibit the kind of disciplinary self-awareness by which we have seen early critical theory defined. Deconstructive critique manifests as a method's reflection upon its reconfiguring, and reconfigurable, attributes.[6]

By positing a critical value to Derrida's deconstruction, the tensions indicated in the preceding chapter between Derrida and Habermas on this matter require acknowledgment. In *The Philosophical Discourse of Modernity* (1987 [1985]), Habermas argues that Derrida's method does not satisfy the requirements of philosophical argument, nor of critique:

> If, following Derrida's recommendation, philosophical thinking were to be relieved of the duty of solving problems and shifted over to the function of literary criticism, it would be robbed not merely of its seriousness, but of its productivity. . . . Whoever transposes the radical critique of reason into the

> domain of rhetoric in order to blunt the paradox of self-referentiality, also dulls the sword of the critique of reason itself. (Habermas 1987, 210)

Derrida's response to the accusation that his deconstruction lacks the requirements of a philosophical social-problem-solving purpose is found in a footnote of *Limited Inc.*, in which he argues that Habermas has criticized his work without actually reading it correctly.[7] For Derrida this claim is evidenced in noting that Habermas has argued against him "without citing or giving a reference for twenty-five pages" (Derrida 1977, 157). Embedded within Derrida's response is the point that Habermas perpetuates a "classical philosophical ethics," in which closely read critique is often absent (1977, 157). We might notice a certain intersection here between Derrida's focus on textual analysis, and the earlier demands of New Criticism. This correlation would not surprise Habermas, given his accusation that Derrida blurs the line between literary study and philosophy.[8]

The ideological hostility between Habermas and Derrida was eventually to be overcome, indeed reaching a point of mutual respect.[9] Regardless, even if one grants aspects of Habermas's critique, it is important that we recognize Derrida's intention to contest the possibility of a signified, established presence. Via différance, we are provided with an insight into the probable destabilization of all structural relations and presentings. This contestation to the transcendent signification of any supposed presence has ramifications for the installation of a "transcendent, natural, real time" referent that is implicit to the social construction time condition.

This destabilization is never more evident than in Derrida's observation that, because difference and différance sound identical in French, a secret collusion occurs between them. Derrida describes différance in exactly these terms in *Speech and Phenomena* (1973 [1967]), noting that "the silent writing" of différance designates "differing *both* as spacing/temporalizing and as the movement that structures every dissociation" (Derrida 1973, 130; author's original emphasis). This characterization of différance, as a temporalization that is a "silence" reiterates the ethereal, hidden, mysterious conception of time that we have encountered. The perpetually deferring temporality of différance is described as a secretive, invisible function, even by Derrida, whereby as reiterated in *Margins of Philosophy*, "the *a* of *différance* is not heard; it remains silent, secret and discreet as a tomb" (Derrida 1982, 4; author's original emphasis).

As considered earlier, if the differential construction of every individuation is this kind of continual deferral, an endless never-arriving/presenting, how words within a linguistic system, gender in a sexed system, or body modifications in a corporeal structure manifest at all seems problematic. In terms of the lateness theme of this book, could this "accomplishing" of signification now be described as perpetually late, if such signification is always deferring elsewhere before it fully arrives/presents? Derrida's answer to the question of how anything manifests is that there are hierarchized binaries, economies of valuation, woven through a structure to "fix" meanings. These binary oppositions consist of a privileged, "inside" term, and an excluded, "outside" term. The sign normalizes the inside term as essential to it and displaces what is not, installing an apparent presence as the sign's or thing's essential quality "to be put in the place of the thing itself, *the present thing*" (Derrida 1982, 9; author's original emphasis).

Derrida questions this dominating role afforded to presence, arguing that meaning is instead unstable. That the signified concept is never present in-itself, but rather relationally relies upon its "absent" counterpart(s), means that it is not only what overtly emerges/presents that produces signification/meaning. Instead, the apparent "absence" to any thing also participates in/as "presenting" that thing. In this light, the earlier discussed prioritization of mechanical time over natural and biological time, or vice versa, seems fragile, as do models that, as Barbara Adam notes, position "modern over traditional time, and commodity over event-based time" (Adam 1990, 153). If according to deconstruction's critique, nothing can be reduced to an either|or proposition, then the distinction of essentially real and natural time, from contingently representational and social times, will come under much scrutiny.

This contestation to the hierarchization of presence over absence encourages me to revisit Augustine's time model. Time, as we have seen for Augustine, is present-centric. The present is everything, whereby past and future do not exist without emerging through a present-as-passing. Here we can observe a point of difference between the conceptions of time proposed by Derrida and Augustine (and indeed, correlatively, between Derrida and the Aristotelean "now"). For Derrida, the present is never in-itself, but rather is only present because what is (seemingly) absent is bound up in the "presenting" of presence. Conversely for Augustine, while past and future cannot exist without a present with which they are co-implicated as duration, there is also a durationless present existing independently from the absence of past or future (Augustine 1961, 266). This aspect of Augustine's argument informs the assumed hierarchization of presence over absence with which Derrida's deconstruction is concerned, as well as the social constructionist time condition that posits a real, universal time as inaccessibly present in-itself. We have engaged another conception in which a vernacular of time's present-centrism is apparent via Mead. However, Mead's demand that there is no time "in-itself" exemplifies an important consistency with the timing of Derridian deconstruction that Augustine's conception lacks.[10]

Contrary to Habermas's critique that Derridian deconstruction lacks an awareness of what is required in philosophical endeavor, Derrida seemingly uses deconstruction's appraisal of the presence|absence binary as a way to reflect upon the responsibilities of philosophy. By critically intervening into what he believes are structuralism's hierarchizations—and acknowledging what is productive rather than merely passive about marginalized states, entities, and demographics—philosophy is asked by Derrida to not simply reverse the binary and prioritize absence over presence. Rather, it is demanded that philosophy calls into question the politics perpetuating hierarchical oppositions generally. Here I argue that deconstruction is a critical reflection on what has been neglected in the present in which a philosopher is embedded. In critiquing the conceptual suppression of absents from presents, it is not that the present should be marginalized, or that the absent becomes prioritized, but "rather that *différance* maintains our relationship with that which . . . exceeds the alternative of presence and absence" (Derrida 1982, 20; author's original emphasis). This appreciation of the instability of structural temporalities is, I would argue, for Derrida a philosopher's methodological responsibility.

It can be recalled from chapters 1 and 2 that social time structures perceived to be more event-centric were conceptually aligned with the materially spatial present. By lacking strict lateness protocols, such social time structures were seen not to have developed far from a natural, ecological, state of time in comparison to social structures with more stringent, mechanical regulations of social time. Through Doreen Massey's application of Derridian deconstruction, however, we can now place this perspective within a history of theoretical subordination of what is natural or materially spatial. Massey argues that with any such dichotomy what emerges is "a presence and an absence; a dualism which takes the classic form of A|not-A" (Massey 1994, 255). If "A" is prioritized, then the other term, "not-A," takes on a character of relative lack. As will be seen, Massey is one of a number of commentators who believe time is conceptually hierarchized over space in that it is "time which is conceived of as in the position of 'A,' and space which is 'not-A'" (1994, 256). This kind of reading could inform our understanding of how the social constructionist time condition separates materially spatial, planetary rhythms from social rhythms.

An important condition by which this position of social constructionism opposes real, universal time from representational, social times concerns the plurality of the latter. Social time's multiple manifestations are said to indicate their contingency, differentiated from the necessarily singular, natural time of materially spatial celestial movements. Materially spatial time is conceptualized as stuck in a perpetually regular loop, whereas social time exhibits alterity. In comparing the politics between the two modes of time, this characterization of them I would now suggest is consistent with Massey's insight regarding the conceptual subordination of space, whereby cultural time marks "change, movement, history, dynamism; while space, rather lamely by comparison, is simply the absence of these things" (Massey 1994, 256–57).

I will here caution that according to the presence|absence binary, Massey's conception might appear to be contrary to Derrida's deconstructive perspective of the hierarchization of presence. The seemingly always *present* time—the perpetually stuck/present natural loop—is the conceptually subordinated element according to Massey and other scholars we are about to consider, compared to a multifaceted (and therefore, the relatively absented) social time. However, we could anticipate that Derrida would respond that in hierarchizing social time over natural time, we *make social time the present node* in the binary, installing it as that which comes to *be time*. This would cohere with our everyday movements, in that as I have posited during the introduction, our lived experience with social representations of time is generally something that we take for granted when living with a consciousness that the clock *is* the only time.

Feminist philosophies reflect upon their evaluation of the time of social histories through a time|space binary that aligns the productive potential of time with social, male achievements, defined by history, culture, education, progress, civilization, politics, and reason. Conversely, the natural, material, and spatial poles of these concepts are marginalized as the supposed stasis of what is cyclical, bodily, maternally feminine, and reproductive. Massey acknowledges this structuring in that "space and the feminine are frequently defined in terms of dichotomies in which each of them is most commonly defined as not-A" (Massey 1994, 257). Similarly for Luce Irigaray, much theory polarizes spatial presence and temporal dynamism according to gendered terms whereby "the feminine is experienced as space . . . while the masculine is experienced as time" (Irigaray 1993 [1984], 7). Females are accordingly, it is argued, excluded from the linearly productive achievements of time-as-culture, when discursively and politically they are marginalized according to correlations with physiological processes. Menstrual and maternity cycles amplify and exemplify this, encouraging equations of females with material spatiality (the body). Frieda Forman and Caoran Sowton hence observe a history of thought in which "the dominant myths of western civilization are those of man marching through time on a mythic journey in search of self, while woman remains outside historical time" (Forman and Sowton 1989, x).

The multiplicity of social time indeed manifests in a regard whereby not all of its forms are politically equal. Barbara Adam notes, evoking deconstruction's methodological self-awareness of attending to structural hierarchies, that some times are "privileged and deemed more important than others" (Adam 1995, 94). Time is purchased, distributed, and sold by an industrial, patriarchal frame of reference, whose inside-male|outside-female dichotomy means, as Forman states, that "time is not freedom for women"

(Forman 1985, 27). This productive entrapment in social temporality is reflected in the numerous demands of home and work.[11] It is women who are, as Rita Felski describes, "juggling child care, frantic about their lack of time" (Felski 2000, 20). Similarly in terms of paid work, female subservience to male time is exemplified in claims that women's employment sometimes earns them much less than the male wage for the same labor time. Forman notes that this means that "in a world where time is money, and where money can mean time, women have little of either" (Forman 1985, 28).

Our brief engagement with feminist discourse has observed the presumed exclusion of the female from socially productive time, due to the conceptual correlation of them with material space. However, my reason for taking the discussion in this direction is not that I am concerned with emancipating females from a materially spatial characterization. In terms of our broader investigation into the social constructionist time condition that polarizes naturally material time and social times, if I have any emancipative intention, it is instead to resurrect material space from this purported exteriority from any cultured incursions that supposedly arrive after it in more complicated, politicized forms.

Structuralisms can often fix static-space in opposition to dynamic-time. Ernesto Laclau actually endorses the structuralist conception that "temporality" as "dislocation . . . must be conceived as the exact opposite of space" (Laclau 1990, 41). The supposedly static, positive nature of space is said to be fixed in place by the structural co-presence of spaces. Laclau duly describes "spatiality" as "coexistence within a structure that establishes the positive nature of all its terms" (1990, 69). In contrast to this spatially closed and stationary system, time for Laclau "takes the form of a dynamic which disrupts the predefined terms of any system" (Laclau, quoted in Massey 1992, 68).

We have noted the issues in homogenizing regional time structures, as well as clarified that the point of integrating such homogeneities to this discussion has been to recognize social time's globally myriad forms. As part of this endeavor, it was observed that Filipino, Caribbean, and African, social time structures are often portrayed as lacking relatively strict conceptions of lateness and so are more closely aligned with the event-based temporality of the "material/spatial present." What now becomes apparent is a reason that such time-structures could be seen to be less dynamic or complicated, whereby Massey observes the critical impression that "the spatial, because it lacks dislocation, is devoid of the possibility of politics" (Massey 1992, 68). According to such a reading, a society's relative lack of lateness protocols would not necessarily be seen to symbolize that particular society's cultural preference—as per readings of "cultural relativity"—but rather to indicate that it is stuck in a materially spatial loop and *unable to develop* the possibility of a more refined temporal politics.

In considering how philosophy intervenes where what is materially spatial about time is separated from what is socially structural about time, the reduction of the materially spatial to an apolitical realm of closure must be targeted. Our focus here is whether a philosophy, in reflecting upon its preconditions, offers critical qualities against structures of privilege and marginalization. It should not be forgotten that Derridian deconstruction argues that a thing's privilege only manifests relationally with its co-productive, structuralist other. The notion of a neat division in a dichotomous pairing, such as between time and space, is problematic. While for Derrida, différance is a temporalizing, such time and timing cannot be a present force at the expense of absent space. Whereas the preceding discussions have correlated the dynamism of temporality with sociality, but separated this sociality from spatiality, différance impels me to ask, what if what is social is materially spatial? In engaging this question, we will benefit by complementing our deconstructive considerations with the structuralist interjections of Pierre Bourdieu.

EMBODIED SOCIAL STRUCTURES

When considering structuralist methodologies, Pierre Bourdieu identifies a theoretical tendency to focus on how socially structural force produces individuals. Contrary to this, in *The Logic of Practice* (1990 [1980]), Bourdieu is interested in tracing objective social structures within subjective actions, while also acknowledging a subject's traces in such structures (Bourdieu 1990, 9). Most interestingly for our study on distinctions between what is naturally physical/material and what is social/cultural, in developing a socially and objectively dispersed sense of subjectivity, Bourdieu argues that one's position within a social structure is reflected by their material, *bodily* practice (1990, 9, 10, 57–58, 66–79). The ways an individual is socially influenced to bodily act, and how such bodily actions contribute to a collective construction of being a bodied subject, marks the body for Bourdieu as a site of interaction between an individual's practices and social structures. What consequently bodily emerges is described as *habitus*, a set of structured-and-structuring practices, manifesting as "systems of durable, transposable dispositions, structured structures predisposed to function as structuring structures, that is, as principles which generate and organize practices" (53). These durable, transposable dispositions include all ways of physically being, such as moving, gesturing, talking, and even, I posit, body modifying.

The resulting dispositional templates frame one's belonging to a class of individuals who occupy similar social positions and share the same habitus. Accordingly, Bourdieu describes *fields*, which perpetuate a particular population's bodily practices. Given that one already belongs to a field in order to acquire its practices, such acquisition takes on an objective, natural impression. This is reminiscent I suggest, of the hyper-normalization of certain beautification, body modification processes, for those of a certain social class as reviewed earlier, which for Bourdieu would reflect a field's "particular class of . . . 'reasonable,' 'common-sense,' behaviours" (Bourdieu 1990, 55). The ensuing regularity of habitus derives from how individual behaviors embody the field, and reproduce the field, from which they have been produced.

Dispositions, ways of being bodily in a social field, become the generative basis of structures that are *structuring* in guiding practices and that are concurrently *structured* in that the acquisition of the power to affect objective behavioral practice is bound up in the acquisition of the objective behavioral practices themselves. In considering body modification as an example of such dialectical, embodied practice, one must endorse the appraisal that Christian Klesse contributes, stating that the act of getting tattooed or pierced has the effect of reflecting *and* creating collectivity (Klesse 1999, 22). Body modifications, constitutive of, and constituted by, habitus are signified/produced in terms of their relational role within social contexts that they/habitus produce. This *near*-circularity to such structural production ensures a perpetual potential for plasticity, whereby as commentator Cheleen Mahar notes, "[B]ecause of its mode of development, habitus is never 'fixed'" (Mahar, Harker, and Wilkes 1990, 11). We could say here that modified bodies modify the modifying, structuring social bodies by which they are modified.

The intervention that Bourdieu provides regarding structuralist theory is to recognize a material/physical, corporeal constitution to sociality. Habitus illustrates how the social body produces an individual body that was already involved in the production of the social body/corporeality. There is not a point of time in-itself when a social body/corporeality *preexisted* an individual body/corporeality, or vice versa. Rather, each is constituted by the other, whereby the direction of social-individual causation is not straightforward. Loïc Wacquant, when describing Bourdieu's perspective, accordingly reviews an individual's trajectory in the latter's sociology as "where the past, the present, and the future . . . interpenetrate one another" (Wacquant 1992, 22). This is consistent with my interrogation of Durkheim's sociology, which argues against the supposition of an exclusively forward-moving, time-linear, cause-effect chain dictating the social-individual relation.

The importance of Bourdieu's conception for the development of this chapter is, I believe, found in its the corporeal/material/spatial focus. While a structuralism such as Anthony Giddens's "structuration theory" recognizes the "positioning of the body" (Giddens 1984, xviv) as an important "medium" for the reproduction of social arenas, he is reluctant to attribute agency to the organic matter/substance of the body. This is apparent in Giddens's characterization of how social systems "*mediate* the physical and sensory properties of the human body" (1984, 36; my emphasis). What this neglects is that such mediation could be entirely material, entirely fleshy, instead implying that bodies simply exemplify the mental milieu of social production. Here I would say that the body is conceived as a "tool" of a social agent, whose enabled and structured constitution enacts what Giddens describes as the "transformation of the body into an instrument of acting-in-the-world" (53).

Conversely, I suggest that for Bourdieu, social structures *are bodied*. Social agency is redefined from a separate Cartesian rationality and willful cognition, to an embodied, practical immersion in physically structural relations. This primacy-in-practice means that what is cultured about bodies in Bourdieu's view "is not something that one has, like knowledge that can be brandished, but something that one is" (Bourdieu 1990, 73). Can we say, accordingly, that Nature-as-body and Social-as-culture coincide here? I feel that such a claim *can* be made, given that in Bourdieu's theory, bodies are the concurrent incarnators and incarnations of a shifting social corporeality. Wacquant's sense of Bourdieu's sociological perspective with which this chapter opened, describing the co-constitution of humans and their world as where "human beings make meaningful the world which makes them" (Wacquant 1992, 7), can now seemingly be understood as a bodily co-constitution. Where subjectivity is socially constructed, the body is socially constructed, however not as a passive entity that feels the force of a separate, social milieu. Rather, the body manifests as an immanent articulation of the world that also articulates the world. This contests the notion that flesh or matter is divided from its social environment, destabilizing what Vicki Kirby describes as the "somatophobia of Western metaphysics that renders matter immaterial" (Kirby 1997, 54–55). The positions encountered earlier that define matter or space as closure in contradistinction to time as potentiality are in this perspective dismantled.

My summary of Bourdieu's position would thus be that a theory of one's materially social production posits not only a location within habitus, but that one *is* habitus, a present embodiment that is also a social history and a future. The body can never be a mere reiteration or copy of what has preceded it, given that the past body (be it an individual's body or a social body) only emerges with, rather than anterior to, the present. At least this is how I read Bourdieu's eloquent description of the body as "the presence of the past in this kind of false anticipation of the future" (Bourdieu 1990, 62). "False" is a curious way to describe this transition, but I argue that it refers to how what exists is not simply permanently fixed in a preceding present, able to be anticipated from an objectively separate state of time. Habitus instead, like a cognitive memory trace, ensures the presence of, and indeed the production of, past experiences as a perpetually plastic, materially embodied social trace.

The understanding of habitus as a dynamic social construction, a corporeal memorial production, between the individual body and the objective, social body captures Halbwachs's earlier encountered conception of memory as a social construction (Halbwachs 1992, 46–51, 124, 173, 182, 187). Individuals and their memories participate in and rely upon collective social environments in order to emerge as individual memory incarnations. If the individual body is similarly productively implicated in the structures from which it emerges, then just as mind and memory are blurred across the social milieu for Halbwachs and Durkheim, so it must now be argued that corporeality is dispersed through a social fabric.

Considerable insight can be garnered from this concerning the earlier engaged appraisals of an already established, natural body, that is intruded upon by secondarily arriving, socially contingent, modification practices. Habitus presents an always already bodying structure, produced by all bodily practices (including those categorized as "body modifying"), which develops plastic, yet durable, forms by which it (the structure of practice) and they (the structured/structuring practices) continue. That habitus reproduces and is reproduced by all practices in a field indicates the plasticity of what is considered to be a "body modification practice." From this insight I argue that if the modified body is always already part of a collectively differentiating fabric for future-and-past productions of habitus and/as the body, then the characterization of body modification practices as arriving contingently "late" to the body seems problematic. In the preceding chapter an argument was developed that the human subject is inescapably, productively implicated in the temporality of objective, social rhythms. This current discussion has attributed that "already implicated" participation to the subject's bodily, material practice.

Given that the subject cannot avoid productively participating in such rhythms though, a question I would like to ask is why should the subject-as-materiality have to *do* anything, as in undertake "practice," in order to structurally temporalize? If we are reconfiguring the social constructionist time condition that the timing of what is naturally material/physical is separate from the timing of the social, then simply being material/physical should have the same politically temporalizing reality or power as the social. In exploring this point, we will revisit the aforementioned method through which philosophy seemingly exhibits an alternate awareness of temporalizing ontologies: Derridian deconstruction. Such navigations will expand on the developing considerations concerning the specific context of body modification, regarding whether social production is a late incursion to, and divergent representation of, a naturally occurring phenomenon.

THE BODY, SPEECH, AND WRITING: THE DERRIDIAN CRITIQUE OF PHONOLOGISM

In considering how Derrida's deconstruction reflects on the natural|social time distinction, we can incorporate his call for an end to the "book" and the beginning of "writing." The book for Derrida is *logocentrism*, or more specifically, as Niall Lucy clarifies, the belief that "before everything else (history, knowledge, consciousness, etc.) . . . there is presence . . . the Logos" (Lucy 2004, 71). This assumption of an untouchably anterior, positively present, origin of truth/meaning, has significant ramifications for two of our considered contexts. Firstly, the notion of the inaccessible, *real* time-source, which social constructions can only contingently represent, would be affirmed by the logocentric view of time as an antecedent, presence in-itself. Secondly, a belief in a logocentric presence for the body underpins the aforementioned self-mutilation argument, which presumes the skin to be a naturally passive, pure, or unmarked surface that precedes the cultural inscription of body modification practices. Such inscription, as we have seen, is criticized by medical practitioners but endorsed by body modifiers on the grounds that cultural practices reinvent the body beyond its absolutely biological, natural state. This is the logocentric frame, with its suppositions that culture distorts and mediates what is naturally and anteriorly present, which I believe Derrida critiques when interrogating spoken|written dichotomies in *Of Grammatology* (Derrida 1976).

What is most relevant from this work is Derrida's discussion of *phonocentrism*. Phonocentric perspectives in Derrida's estimation presume that speaking accesses the greatest proximity to natural or essential meaning in that "the order of *natural and universal* signification is produced as spoken language" (Derrida 1976, 11; my emphasis). Conversely, writing, the written signifier, is in Derrida's reading often excluded as that which supplementarily mediates, or deviates from, the spoken signifier. Phonocentrism's debasement of writing as a signifier of a signifier is not a recent development. Since Aristotle's classification of the written word as a symbol of the more innate spoken word,[12] it has been, according to Derrida, Western metaphysics' archetypal "mediation of mediation and a fall into the exteriority of meaning" (1976, 12–13). This provides a foundation, Derrida argues, for Jean-Jacques Rousseau's denigration of writing, in which Rousseau is cited as stating that "languages are made to be spoken, writing serves only as a supplement to speech" (Rousseau, quoted in Derrida 1976, 144). It is from Rousseau that the anthropological perspectives of Claude Lévi-Strauss later draws, evidenced in Lévi-Strauss's declaration that "Rousseau poses the central problem of anthropology, viz. the passage from an unbridled nature to an ordered society" (Lévi-Strauss 1992 [1955], 229).

Derrida believes that a "phonologism" hierarchizing speech as immediate and marginalizing writing as mediated, distanced, and corrupted is evident in the structuralism of Lévi-Strauss (Derrida 1976, 102). It is this attention on phonocentrism's denaturalization of the social/cultural practice of writing that I believe speaks to my concerns about the social constructionist time condition's characterization of social time structures as distanced from what is naturally real about time. Moreover, in terms of what Derrida calls phonologism's linear "denaturations" of writing (1976, 41), I correlatively observe that body modification practices which *rewrite* the body are characterized as de-natural(izing). When appraising body piercing in this light, its jewelries and other objects, as well as the processes and techniques via which the piercings occur, symbolize late arrivals on the scene of the body. Piercings, as synthetic, socially generated intrusions distance the body from its "natural state." This is not the structurally productive body of Bourdieu's theory but a passive body at the mercy of cultural mediation.

Lévi-Strauss believes that he observes this progression from nature to culture that Western society supposedly enforces upon humans. During a series of encounters with the *Nambikwara* tribe from Brazil, he notes that as an outsider he is not permitted to know their names (Lévi-Strauss, quoted in Derrida 1976, 110–11). By encouraging the children to fight among themselves and then to reveal the names of their combatants to him as revenge, Lévi-Strauss learns the names of the entire community (Lévi-Strauss 1992, 111). This breach of tribal law is interpreted by Lévi-Strauss as an insight into the vulnerability of the primitive Nambikwara peace to infiltration from Western civilization, whereby the anthropologist confesses to "violating a virginal space" (1992, 113).

We have earlier noted Hélène Metzger's reconfiguration of chronologies of primitive mentalities, as identified in the anthropology of Lucien Lévy-Bruhl. Derrida also wishes to unsettle impressions of primitivism by arguing that Lévi-Strauss's perspective reflects an ethnocentric tradition where the confession legitimates a Western insight into what is non-Western. This occurs by journalizing and calculating the relation between culture (Western, civilized society) and nature (Other, primitive society). The Nambikwara, as the Other, are depicted, Derrida critiques, as exemplifying a natural, original innocence, the "index to a hidden good Nature" (Derrida 1976, 114–15). This us-and-them model then indicates to what extent humanity has fallen since its original (primitive) state (1976, 115). Readers might note here how this recalls our earlier considerations regarding the relativization of different, global, social times. As social time structures that are deemed to be more ecologically and event driven are associated with the material present instead of with the complexities of clock-based lateness regulations, then Western time structures in which lateness protocols are more rigorously enforced are reckoned to indicate an increased separation from our natural condition. I therefore suggest that lateness, whether as a culturally inscribed inscription that arrives on the body, or as the regulation of social synchronization according to the industrialized clock, is perceived to symbolize how far from Nature humanity has ventured.

In furthering his argument about the violation of a pre-culturalized nature by Western civilizing mediations, Lévi-Strauss recounts the "writing lesson." This involves Lévi-Strauss distributing pencils and paper to these purely "oral people" to observe what they will do with them (Lévi-Strauss 1992, 288). With these new implements the tribespeople "scribble," which, given Lévi-Strauss's assumption of a "people without writing," are interpreted to be acts of pure imitation (1992, 288). The chief in particular, in realizing that these scribbles are meant to possess meaning, is said to mimic the anthropologist, even if he does not understand what that meaning is.

It is my interpretation that the imitation by which Lévi-Strauss characterizes the chief's writing practice bears an inverted resemblance to accusations of cultural mimicry in the body modification community. This is evident in the perceived appropriation of Eastern imagery by Western body modifiers, having their bodies written with tattooed symbols that they do not entirely understand.[13] Just as Lévi-Strauss's chief apparently participates in a writing practice grounded in imitation, rather than in a comprehension of the written symbols, the Western body modifier is interpreted as similarly intending to harness the symbolic effect/power of Eastern symbols, while only rudimentarily knowing their signification.

For Lévi-Strauss, the writing lesson is an exhibition by the chief that "he had allied himself with the white man, and could now share in his secrets" (Lévi-Strauss 1992, 289). The chief, Derrida notes, has understood writing's "role as sign, and the social superiority that it confers" (Derrida 1976, 125). Similarly I again wish to argue here that in a reverse fashion, the Western body modifier identifies the Eastern tattoo as a sign with different powers from those of their own culture.[14] Tattooing's increasing popularity in Western culture gives it a "fashion/fad" status, something that a "return" to the "purity" of Eastern forms *could* be perceived to counter by conveying a more genuine connection with one's body project and also with one's body. The differentiation between the real, and contingent attempts to access that real, implicit to the social constructionist time condition, reemerges in this context.

In framing his intervention into Lévi-Strauss's anthropology, Derrida notes that it troubles Lévi-Strauss that the pre-literate tribal elder grasps that writing is a tool of power, for "the anthropologist understands what he has taught" (Derrida 1976, 122). As a result of the writing lesson he has given to the Nambikwara, Lévi-Strauss reflects on his complicity in corrupting the innocence of what he believes was a previously natural state. Again, the interpretation of a pure, pre-written nature, should recall assumptions of an unmarked body that is intruded upon by violent, body-modifying writing practices. As we will see, I do not disagree with Lévi-Strauss that there is an essential connection between writing and violence. This is also Derrida's position, stating that Lévi-Strauss is not to be challenged when he relates writing to the exercise of violence (1976, 106). What we can suspect though is that this violence occurs at a more intrinsic level than is supposed by phonocentrism, and according to a different temporality than the separation of natural phenomena, from contingent, social arrivals.

We should not forget that we are exploring Derrida's critique of the phonocentric model as part of the current consideration that I have identified concerning whether simply being bodied, simply being material, is the same as being social. The question here is whether, alternatively, being social also requires bodily *practice*. Via Bourdieu I have claimed that body modifications do not represent late arrivals writing the scene of a natural, anterior body, but were already part of the body's inherent, socially structured, constitution. There was never a pre-socially-structured body. Could a similar logic mean that with Derridian deconstruction, it is able to be argued that a pre-written state never existed in the Nambikwara?

This question can be considered via the Nambikwara prohibition on the disclosure of proper names. In Saussure's semiology, which informs Lévi-Strauss's structural anthropology, a proper name only manifests via its differential relation to other proper names. No name is "proper" in-itself, but instead is systemically produced. This resembles our earlier determination that body modifications generate meaning and identity relationally, rather than positively. Derrida, armed with the Saussurean insight, argues that the proper name only manifests via "classification and therefore within a system of differences, within a writing retaining the traces of difference" (Derrida 1976, 109). This is what Derrida refers to as the death of the proper name (1976, 111). Beyond the systemically negative production of proper names, however, Derrida allows us to explore whether violence does not arrive secondarily from outside the Nambikwara via Lévi-Strauss's facilitation of the disclosure of names, or even via the writing lesson. This concerns the broader possibility that an *originary violence already operates as the differentiality* through which the members of the Nambikwara "naturally" co-exist.

The signature of this position of Derrida's is that writing is apparent even in the Nambikwara community, which Lévi-Strauss conversely interprets to be "pre-writing." In this Derridian perspective, writing must be understood as an "archetypal writing," a generalized writing that differentiates everything structurally. Furthermore, this writing is called "violent" because, and this is very important, *violence is also difference*, violence is classification (Derrida 1976, 110). This writing-violence is, for example, traced in the prohibition of the disclosure of the proper names before they are even disclosed (1976,

110–12). Prohibition *differentiates* between "lawful" and "unlawful," and any such differentiation, for Derrida, exhibits the violence of archetypal writing. Lévi-Strauss inadvertently exposes what is inherently violent about "law" according to this Derridian reconfiguration of violence rather than commits a straightforward offense against it. The real violence in Nambikwara society is an originary network of relational differences that characterizes "writing" in its most primordial mode. Rather than subsequently violating a prior, pure, or real integrity, originary violence operates differentially as the condition of all the various Nambikwara ways of being.

Derridian violence has been relatively recently critiqued by Chung-Hsiung Lai, for whom the adoption of the deconstructive method to serve "critically ethical" investigations must also, ironically, be seen to be unethically constituted. According to Lai, Derrida's deconstruction demands that for philosophical criticism to be "ethical and just, it must be unethical and unjust in the first place in what he calls an 'economy of violence'" (Lai 2003, 23). This refers to the structural co-differentiation between deconstructive justice and injustice. Because justice is the violence of an originary writing that occurs with, rather than without, injustice, Lai determines that the just is always compromised or "betrayed" by what is unjust.

My interpretation of Derridian violence is not, however, of the compromising of any position according to its differentiated other. Instead, it is via the violence upon which deconstructive critique reflects that any position becomes at all. My engagement with Derrida is accordingly situated closer to accounts that endorse Derrida's perspective that violence is an inescapable structural component of the differential becoming of political being. Elizabeth Grosz provides one such example, considering how "rather than simply condemning or deploring violence" we should be interested in how "to explore its constitutive role in the establishment of politics, of thought, of knowledge" (Grosz 1998, 190). The originality of my work with Derridian violence now duly emerges in exploring its temporalizing conditioning of modifications of the body.

Regarding body modification, even before such practices "violently" arrive secondarily or late to an already existing, natural corporeality, I would therefore argue that the body expresses violence as the condition of differentiation from other bodies and things. This insight has ramifications for our inquiry into whether material bodies are socially constitutive without having to undertake "practices." Via Bourdieu's structuralism, the body has been conceived as productive of social, structural rhythms. With the intervention and application of Derridian critique, however, it can be argued that in *being bodied*, a structuring quality is inherently attributable to materiality. The body is inescapably, violently structural due to the material/physical differentiation that it conditions. The notion of a sequence in which violence arrives late to an already originated, natural phenomenon, is being reconfigured.

According to a Derridian logic, my argument is thus that the entire body is written, even accommodating the prohibition of certain practices and their verifiable traces. The typical tattoo shop manifesto of "no minors and no facial tattoos" (DeMello 2000, 20) does not prevent the writing of the underage body or of the face. If tattooed bodies manifest in terms of how they relate to, and differ from, other tattooed and non-tattooed bodies, accordingly there is a bodily writing-violence occurring before tattooing arrives like an anthropologist on a possibly unmarked scene. This violence, the originary or archetypal differentiation of bodies, produces bodies via their structural, relational co-constitution. The non-tattooed body is already, and always, tattooed by its implicated relation to a corporeality that conditions the possibility of tattooed bodies, non-tattooed bodies, and bodies generally. My application of Derridian violence here can be positioned with Pheng Cheah's reading of materiality for Derrida. The originary condition of materiality for Derrida, Cheah observes, is "absolute alterity" (Cheah 2010, 70). Materiality does not mark a mere ontic presence in the form of bodily corporealities, it is "not necessarily corporeality" (2010, 77). Rather, just as I argue that the body is already violently differentiating, with or without the corporeal modification that seemingly distances it irreversibly from its natural state, equally for Cheah the alterity of materiality always already "threatens the teleological self-return of the organism as a self-organizing proper body" (77).

Just as the Nambikwara are not introduced to writing by an anthropologist according to Derrida, but in their differentiation via naming and laws already undertake writing, neither, I argue, is there a straightforward sequence in which the body is introduced to writing by an *exclusive* set of socially presented practices. Wherever difference manifests, there is writing. Writing is not the late or subsequent disruption of an originary, natural present, but rather what is originary and always present is writing in general. This writing should not be lamented, as Vicki Kirby observes, as "a loss of the origin that textuality replaces," but recognized as "an original (worldly) writing through whose radical interiority the referent *presents* itself" (Kirby 2011, 46; author's original emphasis). According to the worldly terms that Kirby introduces in her commentary in fact, writing *is* Nature, whereby everything that Lévi-Strauss situates outside the Nambikwara's natural existence is instead operating differentially within/as it (2011, 57). It is as a result of this deconstructive reflection that Derrida demands that the "writing lesson" is not a passage from speech to writing with the "introduction of writing tools," but rather is a movement from one form of writing to another, which "operates within writing in general" (Derrida 1976, 125).

Derrida's unrestricted sense of writing means that readings have consequently long been offered in response that he simply offers a linguistic idealism (Raschke 1979; Gross and Levitt 1994; Sokal 1996a; Rajan 2002). These interpretations are often directed toward Derrida's notorious claim in *Of Grammatology* that "there is no outside-text" (Derrida 1976, 158). Contrary to the claim of a Derridian, linguistic constructionism, however, my analysis is informed by, and situated with, Kirby's contestation that Derrida's "always already" of writing is anything but a linguisticism that transcends, or relegates to erasure, the reality it describes. Instead for Kirby, the Derridian problematic co-situates nature and culture, presence and absence, inside and outside, in a way where by being involved in, and as, nature, culture does not externally fabricate an originary nature (Kirby 2006, 84).

Kirby argues from this that it is contrarily in the "nature of Nature" to write and to be culture(d) (Kirby 2006, 84). Iris Van Der Tuin notes in response that Kirby's novel reading takes her into the realms of new materialism. The suggestion here is that Kirby exhibits tendencies consistent with a new materialist intervention against the impression that "language is led by an original reality, 'out there'" (Van Der Tuin 2011, 286). In feeling a sense of ideological collegiality with Kirby and Van Der Tuin on this matter, my position is thus distinguished from the interpretation Toril Moi offers of Kirby's Derridian/Saussurean perspective as problematically lacking a human (cultured) author/writer. Moi is concerned that without a human consciousness directing signification, "in Kirby's world, language and matter signify by themselves, without speakers, without any kind of human agency" (Moi 2017, 124). This seems to reiterate the kind of assumption that Lévi-Strauss makes, situating cultural capacities outside a nature upon which they subsequently intrude.

Conversely, if we take Derrida's assertion that nothing is without writing-as-differentiation, as the originary, natural condition of being, then the notion of a linguistic sign that is exterior to writing "falls into decay" (Derrida 1976, 14). The "book" to which Derrida refers as having ended is a logocentrism that never actually existed to oversee signification externally. By contesting this logocentrism, the possibility of an already established sign transcendently presiding over the inscription that is enacted by its signifiers is problematized. Similarly this offers a contestation to the social constructionist condition of an inaccessibly real, singularly logocentric, time that governs the contingent, fabricated signifiers of social time. In considering Derrida's meta-analysis of the nature of writing, this challenge to logocentrism marks an understanding that writing is not an exclusively categorized set of empirical, mediated marks, but instead what originarily conditions writing in empirical forms. If language and the sign do not precede writing, then speech, graphic script, and the body are all differentiating forms or "species of writing" (1976, 8). My characterization of the body-as-writer is seemingly something with which Derrida would agree, given his demand that "the most elementary processes within the living cell" are also "a writing" (9).

That body modification practices are typically comprehended as a reworked "authorship" of a subject's body is consistent with the constructionist impression that "writing" is either an author's *re*-presentation of reality, or a late, contingently social interference of the body well after its natural origination. Now, however, we have corporeal writing. Body modifications exhibit the always writing of/as the material body rather than *introducing* writing to materiality. Just as Derrida's writing-as-differentiality spelled death for the proper name, similarly for body modifications, this incarnated, originary violence-as-writing seemingly contradicts the idea that "body modification" is restricted to a proper, distinct, exclusively socially categorized set of practices. If the originary, natural, and entirely normative capacity of corporeality is that it writes/temporalizes/modifies, then this not only blurs the nature|culture divide but also characterizes the primordial modifier, temporalizer, and writer, simply as the body-subject.

This deconstructive distribution of writing will in the next two chapters switch to considerations of how phenomenological reflections on phenomenological methods open the question of the timing and materiality of theory. What will be at stake in such a debate, as has been occurring in this chapter, is a reconception of a natural logos presiding transcendently and singularly over the temporalities of its secondarily social mediations.

CRITICAL EMANCIPATIONS: DERRIDA OR LEVINAS?

As reviewed at the outset of this book, according to definitions such as those provided by Horkheimer, critical theory is often concerned with practical interventions that can contribute to the "emancipation," if necessary, of groups experiencing oppression/marginalization. Our inquiries have moved through different forms of critique since then, including that of Habermas, as well as of the field of New Criticism. However, in considering Horkheimer's imperative, I will now suggest that if one's natural state, one's material embodiment, makes one a body modifier by sheer incarnation, we should not attempt to liberate certain body modification practitioners and practices from the marginalized, non-normative domains by which we have seen them accused of de-naturally destroying the body. Instead, what is more congruent with our developing insights is a universal emancipation from the categorical restrictions in which only certain practices are identified as "body modifying." Doing this will complement the developing deconstructive reflection on the now reconfigured sequence that characterized body modification as a late, socially informed, intrusion into an already established nature.

The originality of my argument will be illustrated by distinguishing it from that contributed by established body modification commentary. Here I primarily refer to the influential work of Nikki Sullivan, which challenges the interpretation that modification practices bring "to the surface" an individual's preestablished, inner truth. Consistent with my inquiry, Sullivan is interested in problematizing the discourses of self-authorship that are typically associated with body modification. As Sullivan states, the inscription of body modification becomes the codification of *social* excitations rather than the intentionality of a purely self-imposed process (Sullivan 1995, 146). Sullivan's focus is on what body modifications *do*, rather than what they *mean*, demanding that there is no demarcated subject represented in body modification (Sullivan 2009). Rather, the subject/body only emerges via its social relation with other subjects/bodies (Sullivan 2001, 35).

Our engagement with Derridian writing-violence has acknowledged an originary differentiating process that conditions the possibility of distinguishable bodies. Taking a different approach, in *Tattooed Bodies: Subjectivity, Textuality, Ethics, and Pleasure* (2001), Sullivan utilizes the philosophy of alterity of Emmanuel Levinas. Much has been written on the relations and tensions between Derridian and Levinasian notions of difference, alterity, and trace,[15] and my intention is not to contribute considerably to this discussion. Rather, I am interested in how Sullivan's application of Levinasian alterity to an interrogation of body modification practices represents a different set of concerns from the current engagement with Derridian violence, and why these differences are important regarding the context of a critical evaluation of body modification temporalities.

In considering subjectivity, Sullivan engages Levinas's notion of alterity as conditioning the possibility for both the "I" and the "Other" to exist. For Levinas, the I is to some extent separate from the Other but is not autonomous in that its separateness is only possible because the Other exists from which the I is differentiated (Levinas 1969 [1961], 37). Subjects only manifest via structural conditions, and their primordial alterity from that which they are not (but in which they are constitutively "involved") (1969, 35–40). Sullivan's use of this logic of Levinasian alterity exhibits conditional similarities with my engagement of deconstruction's originary violence in terms of observing an ontologizing that differentially conditions all possible identities. Here Sullivan employs a characterization of the primordiality of Levinas's alterity as "a structural possibility that precedes and makes possible" (Sullivan 2004). I am cautious about subscribing to this interpretation that alterity "precedes" possibility, as it appears to position alterity as a preexisting or transcendent ontological mold from which possible subjectivities emerge. Instead, as with Derrida's violence-as-difference, I argue that alterity perpetually (re-)emerges with/as subjectivity-as-possibility, *as* its temporality. Levinas's characterization of alterity is consistent with this if we observe that alterity is only possible if, and as, a subject is structurally, differentially, manifested with its Other. Despite this friction between Sullivan's reading and mine, in contesting the presumption that body modifications represent one's internal, sovereign meaning, Sullivan insightfully argues that the subject does not exist prior to a relational production with the Other, in that "the self exists through and for the Other . . . in and through alterity" (Sullivan 2001, 103). As with Derridian violence, the One/Other co-production is a relational, structural porosity, without permanently establishing preexisting, positive identities.

Concerning the earlier observed accusation that Western tattooing appropriates Eastern imagery in order to harness the symbolic power of the Other culture, what Levinasian alterity *and* Derridian violence clarify is that West and East do not exist in isolation from, or prior to, each other. Rather, West and East emerge concurrently and structurally, the same applying for the civilized and primitive cultures of Lévi-Strauss's anthropology. What seems to be a straightforward adoption of Eastern imagery, we can reconceive as the structural trace of the East that is already operating in the (originary) possibility of Western aesthetics. This can also speak to earlier chapters' comparisons of social time structures that are interpreted to be present-centric versus those that exhibit seemingly more stringent lateness protocols. As has been observed, no time structure can be identified as significant in and of itself. Deconstructive philosophy here provides cautionary insights for how social time structures that function beyond an investigator's default impression of time are framed.

Similarly, the trace conditions the (co-)production of bodies, or as Sullivan notes in Levinas's terms, as "having-the-other-in-one's-skin" (Levinas 1998 [1978], 114–15). The subject never comes to the Other predefined but is perpetually, corporeally produced, by a beyond that it equally constitutes. For the modified body, meaning becomes a tenuously blurred rather than a reliably self-expressive, exercise. This acknowledgment is a strength of Sullivan's work, who concludes in terms of Levinasian alterity that "not only is the distinction between self and other undermined by Levinas, but the question of what the tattooed body of the other means, or whom the tattooed wo/man is, is rendered redundant" (Sullivan 2001, 111).

However, Sullivan's argument stops at the point where my current inquiry overflows into the *categorization* of body modification practices. The redundancy of the notion of in-itself meaning/identity for such bodies is, I argue, bound up with the equally redundant notion that any practice is essentially, in-itself, body-modifying. Practices can only be acknowledged as modifying due to the originary materiality that conditions the differentiation of bodies. Sullivan rightly notes that originary alterity challenges the "assumption that meaning/identity is reducible to an essence present in the textual body of the other" (Sullivan 2001, 111). I extend such a thesis though by arguing that without characterizing each body modification practice in equally redundant terms, the transcendent preexistence of modifying practices to modified bodies is assumed.

This argument is similar to how I earlier approached a reconfiguration of Bourdieu's structuralism according to the deconstructive method. For Bourdieu, transposable dispositions, ways of bodily being in the world such as moving, gesturing, and talking constitute the corporeal/material production of material-social structures that concurrently produce corporeality. However, this chapter has argued that what primarily conditions these *ways* of bodily being in the world is an ontology in which simply *being* bodied differentiates/produces/temporalizes materiality/spatiality. Sullivan is consistent with Bourdieu in arguing against a defiantly subjective self-production. As with my Bourdieu discussion though, I deconstruct this point further by recognizing the material/spatial body as the primordial modifier/producer/temporalizer, which conditions Sullivan's ways of body modifying, just as it conditions Bourdieu's ways of bodily being.

This difference is subtle but crucial in recognizing the different stakes at play when deconstructive philosophy critiques the temporalities conditioning both philosophical, and socially scientific, assumptions. If no bodies preexist alterity's relational conditions, whereby tattooed bodies only manifest with/as other bodies and entities, then the implicated producer, tattooing-as-writing, tattooing-as-time, must manifest concurrently with such bodies. There is not a preexisting act that humans undertake or employ that was permanently defined in a distant, inalterable past as "body modifying." Rather, just as we have considered how the past becomes the past that it is, in and with the present, something like tattooing only becomes a body modifying practice during a material encounter when tattooed bodies, as time, manifest. Tattooing cannot be permanently categorized as introducing modification (writing) to the body, because tattooing has not even become body modifying without originary, modifying bodies, that differentially co-condition the possibility of tattooed and "non"-tattooed bodies.

This reflection on body modification discourse also provides an insight into the timing of the critical method, where I want to argue that the categorization of critique only manifests with, and as, the bodies of work being critiqued. Just as body modification practices only become body modifying via the originary, material alterity of differentiated bodies, equally the practices of critique only become critical via the differentiation of theories. The distinguishing feature between the realm of critique and that of body modification is the materiality of the latter. However, building on how writing has been deconstructively conceived in materializing terms in the preceding discussion, in the coming chapters the context of the materialization of theorization will emerge more prominently.

The primary role of Derridian violence in the argument of this chapter, and of Levinasian alterity in Sullivan's work, illustrates a key methodological difference between Sullivan's focus and mine. Both Derridian violence and Levinasian alterity contest notions of demarcated, exclusively individual, subjectivity. In terms of the specific context though—body modification practices—the efficacy of Derridian violence emerges. Derrida's rewriting of violence beyond the reductive dichotomies of "good" and "bad," "before" and "after," or "natural" and "cultural" acknowledges violence as originary and natural. As a result, I believe the most effective way to critique and contest instilled discourses that, as observed, condemn non-normative body modifications because of their denaturalizing, late-arriving, "corporeal violence," is via Derrida's reconfiguration of "violence." Furthermore, this avoids futile attempts to rescue particular body modification practices from their non-normatively "violent" characterizations by instead focusing upon the primordial and productive conditions of all violence. This is the strength of the Derridian argument in critiquing body modification discourses. Derrida embraces "violence" by rewriting it, whereby what is violent is not a socially informed disruption of a pre-originated, separately natural, incarnation. Violence, rather, is that which *is* incarnation.

Contrarily, Sullivan's deployment of Levinasian alterity does not deconstruct "violence" but instead divorces violence from alterity. Sullivan denounces violence, characterizing it as a subject's domination of the Other that denies any becoming/alterity with the Other. In distinguishing violence from alterity, Sullivan claims that "such a *disavowel of alterity* results in an hegemony of the Same that *is tantamount to an act of violence*, to a single blow in and through which 'I' become master" (Sullivan 2001, 139; my emphasis). Sullivan's interpretation that Levinasian alterity is inconsistent with violence again emerges with the claim that "textual violence does not consist of marking and being marked, but rather is the result of disavowing such a process . . . that reduces the Other to the Same" (2001, 134).

I interpret alterity as conversely being more congruent with violence. Derridian, originary violence demands that "textual violence" *does* indeed consist of the relational marking and production of material bodies, by material bodies, contrary to Sullivan's critique. This violence-as-time is an originary production that never "disavows alterity," nor "reduces to Same," as Sullivan further claims but rather conditions alterity's incarnating/worlding of world. Sullivan's exclusion of violence from alterity's production of differentiated bodies is an issue, in my estimation, when we approach any recalibration of characterizations of body modification as violently denaturalizing. By accommodating violence within alterity's conditioning, the destabilization of suppositions of the late or subsequent arrival of a violence which *de-natures* the body becomes possible.

It is from this development that deconstruction has framed a contestation to the characterization of body modification as a separately late, social incursion, into a natural scene. But are we now in a position where all social phenomena are natural phenomena? This seems like a particularly contentious claim to make when considering a context such as the possible effects on global ecologies of human, social, industrial development. In response a reader could be concerned about how we view the social and cultural productions from which climatic changes might result, if the social and the natural are not deemed deconstructively to have separate origins. We will open such a direction via phenomenological critiques of presuppositions of natural and social timings.

Chapter Five

Material Climates, Material Theories

A Late Response or a Self-Reflection?

> There is no problem of the *alter ego* because it is not *I* who sees, not *he* who sees, because an anonymous visibility inhabits both of us, a vision in general, in virtue of that primordial property that belongs to the flesh, being here and now, of radiating everywhere and forever, being an individual, of being also a dimension and a universal.
>
> —Maurice Merleau-Ponty, *The Visible and the Invisible* (1968 [1964], 142)

MERLEAU-PONTY'S CRITICAL, PHENOMENOLOGICAL BODIES

By comparing social time structures in which lateness is a significant factor for social synchronization, with social time structures in which lateness is reported to be less regulated, the social constructionist time condition's impression of social time's contingent plurality has emerged. Furthermore, in positing that social time structures are subsequent representations of an inaccessible, natural, real time, a hierarchy regarding the politics of social times has become apparent. Consistent with accounts of humanity's escape from natural origins and to a civilized state, the stricter structuring of temporality has been reported to indicate a developmental ascendency. Time is natural, but social time is civil. If a time structure lacks exact protocols, controls, and regulations regarding lateness, then its correlative association with the ecological, event-based, rhythms of the "material present" has conceptualized it as less distanced from an original, natural time state than its more "tardiness-strict," global counterparts.

It became further apparent, during last chapter's journey into the context of body modification discourses, how lateness presents as an informative parameter when comparing conceptions of secondariness between natural and social timings. Via a deconstructive reflection of structuralist methods and preconditions, the neat distinction between naturally material and socially mediated originations has been destabilized. In now broadening the context from the localized materialities of bodies to the globalized materiality of planet Earth, we are going to ask whether this destabilization of the social constructionist time condition is apparent in phenomenological meta-reflections on human perceptual relations with worldly reality.

These considerations will be applied to discourses within the environmental humanities, media, and other sources that feature a politics of secondariness about a human, socialized, industrialized intrusion on a naturally preexisting, worldly environment. This marks the third context of this project: the concern that collective human responses and interventions to the consequent changes in the environment and climate are occurring *too late* to guarantee our survival as a species. The "late" theme is thus again relevant according to the double meaning of it that we have encountered previously, in speaking both to a sense of secondariness and also to the notion of potential mortality. By considering the culturally constructed human perception of this changing planetary reality, this discussion also incorporates the question of the timing of theoretical inquiry. This, the fourth flagged context of this work, asks whether cultural production, theory, arrives representationally late on the scene of an already originated, natural reality into which it inquires. Conceptions of theory, materiality, and temporality here collide.

The phenomenology of Maurice Merleau-Ponty will be our primary engagement on these matters, in particular his attention on the embodied reality of experience. While a general investment will be felt in this chapter's considerations by environmentally conscious readers from all backgrounds, for students and scholars of the phenomenological philosophies, there will be the added benefit of encountering novel applications of Merleau-Ponty to questions of the timing of human-climate, and theory-reality, relations. We will first unpack the embodied aspect of his thought in relation to the phenomenological method that precedes and informs his position before considering what I believe is "critical" about Merleau-Ponty's method. We can then apply his reconfiguration of the phenomenological method to an intervention into the aforementioned climatic and theoretical contexts, where I identify the presence of the social constructionist time condition.

Our considerations of Merleau-Ponty should firstly observe that it is when appraising René Descartes's doubt of the world's existence, attributable to Descartes's musing that when dreaming we do not realize that such perceptions are not real, that Merleau-Ponty counters that skeptical arguments rely on a secondary perception (the dream) to question a primary, direct, experiential perception. That primary perception occurs from having already been in a world about which can then be dreamed. For Merleau-

Ponty, the Cartesian dream "argument postulates the world . . . the *true in itself*," but that the world is "*secretly* invoked in order to disqualify our perceptions" (Merleau-Ponty 1968, 5; my emphasis). Conversely, Merleau-Ponty does not regard inquiries concerning perception as potential contestations to the world's possible existence but instead as facilitating considerations of "what it is for it to exist" (1968, 96). This marks why this study of the constructionist separation between natural, "real time," and human socially constructed time, will benefit from engaging Merleau-Ponty. Skeptical arguments presuppose a thing *in itself*, oppositionally *outside* the human, about which the truth cannot be entirely known. Merleau-Ponty, though, interrogates the very presumption of a thing in-itself, presenting a congruence with our interrogation of the conditions by which an in-itself, real time-source is posited as being inaccessible to supposedly distorted, human conceptions of it. It is via this interrogation of the in-itself that I will soon locate the critical value of Merleau-Ponty's work.

The notion of the in-itself is interrogated by Merleau-Ponty in exploring the singularity of perceptual relations between observer and observed. Perceptual relations are conditioned by the embodied, spatial positioning between an observer/perceiver and an object being observed/perceived. It is in the text from which the above citations are drawn, *The Visible and the Invisible*, that Merleau-Ponty duly describes how a "table before me sustains a singular relation with my eyes and my body" (Merleau-Ponty 1968, 7). We will explore this notion of singularity throughout this chapter, a condition of which is its reference to what is common between the observer/perceiver and the observed/perceived. The important point to take now is that because bodily movement, or physical positioning, beyond mere eyesight alters the perspective of this singular perception, there is a corporeal, material, component to experience (1968, 8). This alone does not mean perception is entirely material/physical but rather that corporeality *conditions* perception.

The significance of Merleau-Ponty to this corporeal, material focus can be situated by distinguishing his work from the ontological models offered by two phenomenological giants, Edmund Husserl and Martin Heidegger. Adjacent to these phenomenological reflections on phenomenological methods, it is also worth recalling Derrida's earlier contestation to the prioritization afforded to presence. Derrida is generally critical of phenomenology, exemplified in *Speech and Phenomena* in which he argues that phenomenology can only ever *present* a logocentric metaphysics.[1] While exploring Merleau-Ponty, a response to this claim from Derrida will allow us examine the constructionist time condition's distinction between natural time and social time according to a presence|absence binary.

Merleau-Ponty's phenomenology continues a line of inquiry generated by Edmund Husserl. Husserl posits that the existence of the world is generally taken for granted, whereby such a belief is held indifferently. What Husserl calls the "natural attitude," which is an indifference about what one believes, cannot suffice, he suggests, demanding a phenomenological reduction (*epoché*) in which judgments about the existence of the external world are suspended until an interrogation of one's consciousness can justify such belief. This challenge to indifference is consistent with an aspect of critical theory that we have encountered in the Frankfurt School, Habermas, and even in New Criticism. Such a feature of Husserlian phenomenology will reemerge when we discern the ways that Merleau-Ponty interrogates the social present. In Husserl's specific focus though on the *perceptual* present, in *Ideas I* (1983 [1913]) a meta-commentary is offered targeting everyday attitudes around belief, as well as the false premises that philosophical methods assume:

> *We put out of action the general positing which belongs to the natural attitude*; we parenthesize everything which that positing encompasses with respect to being: *thus the whole natural world* . . . I am not negating this "world" as though I were a solipsist; I am not doubting its factual being as though I were a skeptic; rather I am exercising the "phenomenological" ἐποχή which also completely shuts me off from any judgment about spatiotemporal factual being. (Husserl 1983, 61; author's original emphasis)

The phenomenological reduction does not negate consciousness of the world. The world is still actual, "it is still there, like the parenthesized in the parentheses" (Husserl 1983, 9). However, it does require that we bracket (parenthesize), or do not action, our judgment concerning the world's actuality. By suspending judgment, all that remains in one's consciousness is the world's givenness, its sheer experiential facticity. Consciousness of worldly phenomena is hence proposed to be absolutely primordial, as "the infinite world of absolute mental processes—the fundamental field of phenomenology" (1983, 114).

You might be thinking that the apparent *transcendence* of this absolute consciousness seems inconsistent with our argument developed in chapter 3 concerning subjective time. Furthermore this separation of the worldly real from the human judgment (representation) of that real seems to simply reiterate the natural-time|social-time split. We should remember though that we are engaging Husserl in order to contextualize Merleau-Ponty's phenomenological interventions. While Merleau-Ponty is indebted to Husserl's project, his own inquiry in *Phenomenology of Perception* (Merleau-Ponty 1962 [1945]) demands the impossibility of such a reduction. Merleau-Ponty's claim of impossibility is the first condition that will problematize the Derridean critique that phenomenological methods install a pure, hierarchical present. As noted, according to Merleau-Ponty our worldly existence is "primordial" (primitive) and "direct." The phenomenological reduction, Merleau-Ponty discerns, simply makes us aware of this:

> It is because we are through and through compounded of relationships with the world that for us the only way to become aware of the fact is to suspend the resultant activity . . . to put it "out of play" . . . break with our familiar acceptance of it. (Merleau-Ponty 1962, xiv–xv)

We might become aware of our directly experiential worldly relations through the phenomenological method. However, this does not mean that we can entirely break from these worldly relations in order to reflect upon worldly experience from a representational position of without. This is what I take from Merleau-Ponty's point that the experiential conditions of being-in-the-world (*être au monde*) prevent the *complete* reduction of experiential phenomena.

I will here compare that while, according to the social constructionist time condition, we are unable to break with our socialized state in order to reflect upon time in its inaccessible, naturally worldliness, for Merleau-Ponty there is something about our being that is always in its worldliness and from which we cannot entirely break to reflect upon it. What is socially inescapable for the social constructionist position we are engaging is for Merleau-Ponty a perceptual and experiential inescapability. Any reflection upon experience is already a being-in-the-world experience, meaning for Merleau-Ponty that "radical reflection amounts to a consciousness of its own dependence on an unreflective life" (Merleau-Ponty 1962, xvi). This description of life as "unreflective" is not straightforward, and my argument will unpack its complexity later. Nevertheless, what I ask us to take from this passage now is that for Merleau-Ponty, one cannot parenthesize being-in-the-world nor put it out of play by remaining transcendently present to the self without worldly experience. This is also the foundation of Merleau-Ponty's earlier critique of Descartes. The reduction is just one of the many modes of being-in-the-world and is irreducible. This prevents the phenomenological reduction from "completeness":

> The most important lesson which the reduction teaches us is the impossibility of a complete reduction . . . . If we were absolute mind, the reduction would present no problem. But since, on the contrary, we are in the world, since indeed our reflections are carried out in the temporal flux onto which we are trying to seize . . . there is no thought which embraces all our thought. (Merleau-Ponty 1962, xv)

This catch, of "being-in-the-world" to which the phenomenological reduction directs its attention *while* being of *that same* world means that the immediacy of experience, the "things themselves," are never merely reducible to a presence in-itself to which one can reflectively return. Every subject is always implicated in the world that constitutes their consciousness of the world, whereby there is never a world to be reflected upon that is an immediate in-itself, outside the subject. Conversely, social constructionism does posit an in-itself world outside not only individual subjects but also outside collective populations.

In considering Merleau-Ponty's reflections on the Husserlian phenomenological method, it is not that Merleau-Ponty entirely refutes the phenomenological reduction. Rather, he challenges Husserl's characterization of it as a *transcendental* consciousness. A contestation to the notion of transcendence is notable given that my interest in the constructionist separation of natural time from social time is that it installs a condition of transcendent inaccessibility between the time of the real world and the time of which human populations are collectively, socially conscious. Rather than supposing a possibly transcendent logic, Merleau-Ponty is instead intrigued by the way the reduction makes "unreflective," primordial being-in-the-world, apparent. The reflection that conditions the phenomenological reduction

> does not withdraw from the world towards the unity of consciousness as the world's basis. . . . [I]t slackens the intentional threads which attach us to the world, and thus brings them to our notice; it, alone, is consciousness of the world. (Merleau-Ponty 1962, xv)

Without the reflection of the reduction, Merleau-Ponty above believes our phenomenological being-in-the-world would go unrecognized. Husserl actually also posits this, where if "the phenomenological attitude had not been recognized . . . the phenomenological world had to remain unknown" (Husserl 1983, 66). However, something requires attention. Here, "recognition" of being-in-the-world is *consciousness* of being-in-the-world. When Merleau-Ponty argues that consciousness is not a transcendent "withdrawal" toward absolute consciousness, "consciousness of the world" is instead defined as a worldly *immanent*, directly perceptual experience. Yet as has been highlighted, in this early era of Merleau-Ponty's phenomenology, we have just encountered the description of being-in-the-world as our "*un*reflective life" (Merleau-Ponty 1962, xvi; my emphasis). This notion of "un" must be approached carefully, in order that our appreciation of it does not contradict the characterization of consciousness as always being of the same world upon which it reflects. Indeed, I am concerned that the "un" of reflection might even conjure notions of social constructionist inaccessibility, or in the context of Merleau-Ponty's body-centric discussion, attribute the body with passively *un*thinking qualities.

Congruent with our earlier interrogation of suppositions of transcendent relations between the natural world and socially constructed consciousness, Merleau-Ponty problematizes Husserlian transcendental idealism. Merleau-Ponty describes being-in-the-world as "doing away with any kind of idealism," and instead being an immanently relational consciousness, "a dimension in relation to which I am constantly situating myself" (Merleau-Ponty 1962, xiv). Here Merleau-Ponty coheres with our developing inquiry into the supposed secondariness of subjective and collective representations of the world, compared to an anterior world about which is idealized. Because Merleau-Ponty's worldly act of perception, being-in-the-world, manifests a subject that concurrently is of the world, I argue that if the phenomenological reduction is directed toward a world "in its givenness," then being-in-the-world is the primordial phenomenological reduction. This is because, via being-in-the-world's direct experientiality, we already only have a relation to a world "in its givenness." Judgments/reflections about the world supposedly outside that givenness, which Husserl wanted to bracket in order to avoid the "natural attitude," are actually simply modes of always already being-in-the-world in its givenness. Rather than positing this as Husserl's transcendental/absolute idealist/mental processes, the perceptual body is the focus. Let us revisit why Merleau-Ponty's attention to the body is important to our study.

The social constructionist time condition requires that time structures that lack strict protocols regarding lateness are more closely restricted to an association with the natural, celestial, event-based temporality of the "material present." As a critical response to this, complementing the deconstructionist insights from the previous chapter, we are exploring how phenomenology reflects on the presumed secondariness of the representation of time. Merleau-Ponty's focus on how our subjective relation to the "real world" is directly, materially conditioned, rather than inaccessibly representationally framed, begins to counter social constructionist assumptions regarding this theme of secondariness. Moreover, the body, as a site of naturally real and socially constructed intersections, has emerged as a lens through which the relative lateness of representational temporalities can be explored.

As noted by more recent commentaries,[2] however, the body, contrarily, sits tenuously and open for debate in Husserlian phenomenology. *Cartesian Meditations* (Husserl 1999 [1931]) evidences potential key differences between Husserl's conception of the body and that of Merleau-Ponty. For Husserl, perceptual acts are bodily acts. This does not automatically acknowledge the perceiving subject as bodied though. Instead, the body only manifests reflexively, by one organ perceiving another

> *perceptually* active, *I experience all of Nature, including my own animate organism*, which in the process is reflexively related to itself. . . . I "can" perceive one hand "by means of" the other—a procedure in which *the functioning organ must become an Object and the Object a functioning organ.* (Husserl 1999, 97; author's original emphasis)

This intentional bodily reflexivity, the incorporation of tactile sensations by which the body "localizes" and becomes, is observed by Husserl in *Ideas II*

(1990 [1952]) to be conditioned by the double apprehension of touch (Husserl 1990, 155). In touching one can feel oneself feeling. The ramification for subjectivity is that the body is not coincidental with the subject who experiences such sensations. Instead, for Husserl, the body manifests as a possessed "field of localized," tactile sensations, an intermediary between the Ego/subject and the material realm from which sensations are incorporated. Or in Husserl's terms, the subject is "a counter-member of material nature . . . an Ego to which a Body belongs as a field of localization of its sensations" (1990, 159). Whether this means that the body is *between* the perceiving subject and the perceived object, as something *attached to* the subject, occupying the external realm in a way the subject does not, is as already indicated, open for debate. Husserl potentially encourages such readings by describing the body as "a thing 'inserted' between the rest of the material world and the 'subjective sphere'" (169).

This notion of insertion is conditional according to Dermot Moran, however, who has recently argued that while for Husserl "the body is in a sense passively inserted in the world," any interaction we have with others is corporeal, in that "human-beings-are-being-with-others-in-the-common-world through being embodied" (Moran 2017, 42). While there might not be a purely oppositional structure between materiality and subjectivity in Husserl, a concern is whether, if Husserl demands a "Transcendental Ego" to which bodily sensations belong, self-consciousness is positioned as ontologically anterior to the body and bodily experience. I am receptive to this kind of reading of Husserl, given that he states as much in *Cartesian Meditations* in that "the being of the pure ego and his *cogitationes*, as a being that is prior in itself, is antecedent to the natural being of the world" (Husserl 1999, 21; author's original emphasis). Husserl's Transcendental Ego precedes the entire natural realm in fact, whereby "natural being is a realm whose existential status . . . continually presupposes the realm of transcendental being" (1999, 21). Indeed, chapter 1 of Section Two in *Ideas II* extensively (Husserl 1990, 103–27) characterizes the Ego as that which conditions the possibility of the *subsequent* corporeal subjectivity that emerges from the double-apprehending touch (1990, 107).

Despite making distinct the transcendence of conscious experience from the realm of external objects, Husserl does not characterize his phenomenology in exactly Cartesian, dualist terms.[3] Nevertheless, it is due to the corporeal, subject-object synchrony of Merleau-Ponty's latter method that I believe he offers a greater relevance to phenomenological destabilizations of the social constructionist time condition. This is despite Merleau-Ponty's concession in the working notes found in *The Visible and the Invisible* that his earlier text, *Phenomenology of Perception*, begins with a consciousness that is separate from the objects with which it is intertwined (Merleau-Ponty 1968, 200). As Merleau-Ponty's work develops though, commonly corporeal conditions for perceptual experience emerge more prominently than in either the phenomenology of Husserl or even that of Martin Heidegger.

Heidegger's seminal text *Being and Time* (1962 [1927]) focuses on *Dasein*, a study of the relation between general Being and particular beings, which "in its very Being, has this Being as an issue" (Heidegger 1962, 137). However, references to the bodies of beings within *Being and Time* are scarce. Heidegger declares in terms of the directionality of the body that "Dasein's spatialization in its 'bodily nature' is marked out in accordance with these directions (this 'bodily nature' hides a whole problematic of its own, *though we shall not treat it here*)" (1962, 143; my emphasis). Jean-Paul Sartre famously attacks this aspect of Heidegger's work, asserting that "Heidegger does not make the slightest allusion to it [the body] in his existential analytic" (Sartre 2003 [1943], 405).

This critique is later acknowledged by Heidegger in the *Zollikon Seminars* (2001). During these seminars, delivered between 1959 and 1969, Sartre's attack about why Heidegger "only wrote six lines on the body in the whole of *Being and Time*" (Heidegger 2001, 231; author's original emphasis) is raised. Heidegger responds that while being "unable to say more [about the body] at the time" (2001, 231), *Being and Time* assumes that humans could not participate in the "world-openness" of *Dasein* if they were not constituted by "bodily nature" (231). The being-in-the-world (*In-der-Welt-sein*) that *Dasein* explores is, Heidegger demands, "always already fundamentally consisted of a receptive/perceptive relatedness to something which addresses us from out of the openness of our world" (232). Heidegger further describes this openness of the receptive/perceptive human body as not limited by corporeal limits (232–33). Rather, the lived-body (*Leib*) stretches beyond apparent, corporeal boundaries. In accusing Sartre of being blinded by a Galilean conception of the corporeal object (*Körper*) that is restricted by spatial boundaries to being in one place at one time, Heidegger instead notes that while "the corporeal thing stops with the skin," one "cannot determine the phenomenon of the body in relation to its corporeality" (86). There are interesting congruencies and tensions between this argument, and Merleau-Ponty's, that will be reflected upon in the aspects of this chapter dealing with bodily limits. As this unfolds, so the specific applicability of Merleau-Ponty's method to our inquiry in recognizing a directly embodied, rather than a transcendently representational, relation to the world will become apparent.

Why Merleau-Ponty's focus on the phenomenological *body* constitutes a *critical* endeavor can be evaluated, I believe, via his extension of Karl Marx's materialist dialectic. Marx's dialectic itself needs contextualizing in that it responds to the conception of the dialectic that is attributed to Georg Hegel (but that Hegel identifies in the preceding work of Immanuel Kant).[4] The Hegelian dialectic loosely concerns how any thesis induces a negating or contradicting antithesis from which an ideological resolution is reached via a resulting synthesis. In the first volume of *Capital* (1976 [1867]), Marx critiques how Hegel's dialectical syntheses require what is "concrete" about the world to pass through an ideologically abstract construction of it. Given his focus on material conditions, Marx radically opposes his own dialectic to Hegel's:

> My dialectical method is, in its foundations, not only different from the Hegelian, but exactly opposite to it. For Hegel, the process of thinking, which he even transforms into an independent subject, under the name of "the Idea," is the creator of the real world, and the real world is only the external appearance of the idea. With me the reverse is true: the ideal is nothing but the material world reflected in the mind of man, and translated into forms of thought. (Marx 1976, 102)

"Materiality" here is not a reference simply to the physicality or ecology of the world. Rather, it is concerned with how in order to survive, and for human existence to be perpetuated, individuals must enter into relations with each other to produce life's necessities. Despite this broader perspective, the aforementioned ecological and physical aspects of the world are still inherently involved in the laboring designed to serve our ongoing existence. Marx notes of this dual constitution that "the physical bodies of commodities, are combinations of two elements, the material provided by nature, and labour" (Marx 1976, 133–34).

The consequent power relations that historically manifest around the production and exchange of material-necessities-as-commodities are Marx's target. As the bourgeoisie takes control of the means of production of these necessities, for Marx the worker is distanced from their direct, material sustenance-source, given how the "immediate producer, the worker" has "ceased to be bound to the soil" and instead is "the slave or serf of another person" (Marx 1976, 875). Of particular concern are the economic hierarchies that historically manifest from the exploitation of the worker, who has been "robbed of all their own means of production" (1976, 875). Implicit to an attention on this historicist aspect of Marx's dialectical materialism is the associated aspiration of an "end to this history," in which the proletariat working class takes control of the means of the production, leading to the supposed emancipation of all individuals from a state of exploitation.

The materialism of the Marxist critique appeals to Merleau-Ponty. This is despite Merleau-Ponty's contestation, as he describes in *Adventures of the Dialectic* (1973 [1955]), to the "permanent revolution" (Merleau-Ponty 1973, 206) in which Marx supposes history might "resolve." As translator Joseph Bien notes, for Merleau-Ponty "Marxism represented itself as a realization of theory and praxis, of critique and practice," however, having posited an ultimate end point, "it failed in practice in its attempt to present a potentially universal class, the proletariat, which would bring about a reconciliation of all men" (Bien, quoted in Merleau-Ponty 1973, xxviii).

Part of the problem in Merleau-Ponty's estimation seems to be Marx's primary focus on the relationship between the proletariat individual and the production process. Alternatively, and in a manner that opens considerations of the critical characteristics of his phenomenology, Merleau-Ponty is more concerned with the proletariat's individuation as a "historical being" (Merleau-Ponty 1973, 176). Rather than presenting an individual who could be emancipated from historical exploitation by taking control of the means of production, Merleau-Ponty wants to recognize the inescapability of being historically, situationally defined, whereby "the proletariat *is* this philosophical meaning of history" (1973, 45; author's original emphasis). It is this perpetual opening of the individual onto, and as, history, an integral feature of Marx's materially dialectical history, which means for Merleau-Ponty that "what then is obsolete is not the dialectic but the pretension of terminating it in an end of history" (206).

This theme can be expanded into the context of the body's materiality, via Diana Coole's assertion that Merleau-Ponty's critical coherence with Marx's dialectical materialism regards their respective positions on the historical constitution of one's *bodily being*. This becomes apparent for Marx, according to Coole, in what she describes as his "insistence on the historicity of the bodily's sensuality" (Coole 2007, 108). What I interpret in Coole's claim is that rather than sensuality being attributed to something timelessly essential about human physiology, the claim from Marx is that "the forming of the five senses is a labor of the entire history of the world down to the present" (Marx 1973, 109). It is through the historical production of new objects within the "material world" that new modes of human sense-perception develop. The "critical aspect" (Coole 2007, 108) of this historical materialism for Marx, as Coole notes, is identifiable in his commentary on the prevention of our access to some of these perceptual modes. This restriction of perceptual processes is attributable to the bourgeois possession of new objects and their associated properties and production facilities, which over time cuts citizens off from a sensory-rich world of objects-as-commodities (Marx 1973, 109–11).

Marx's model of the historical construction of the material body compels Merleau-Ponty in Coole's reading to "remember that his central ontological figure—the practical, perceiving body—is situated in socioeconomic and historicocultural contexts to which it responds and which transfigure it" (Coole 2007, 108). I am in agreement with Coole, in that by combining a corporeal *and* historical materialism inspired by Marx, Merleau-Ponty goes beyond the applications to existentialist philosophy that are facilitated by considerations about perceptual bodily relations to also offer critical interpretations of our historicized materialization/corporealization. In an often-referenced passage in recent debates,[5] Merleau-Ponty makes this explicit claim in *Sense and Non-Sense* (1964 [1961]), referring to Marx's focus on the histories of human practice as "this concrete thinking, which Marx calls 'critique' to distinguish it from speculative philosophy, is what others propound under the name 'existential philosophy'" (Merleau-Ponty 1964, 133). As Coole notes in reviewing this point, what Merleau-Ponty means here by "concrete" is not merely the physically material, although this is definitely involved, but also that which is embedded in socialized, historicized, ways of living (Coole 2007, 109).[6]

This is what I position as perhaps the pivotal, critical element of Merleau-Ponty's phenomenology: its critique of that which presents as unquestionably, timelessly true/established. Coole would seemingly agree, arguing that we can focus on Merleau-Ponty's phenomenology as a critical theory in terms of how he is "motivated by resistance to reified practices or structures that entail closure" (Coole 2007, 117). The earlier reviewed "in-itself" that the perceptual body politics of Merleau-Ponty's phenomenology *contests* is, in my opinion, the archetype of this ontological "closure." By unsettling the ideological domination of the "in-itself," Merleau-Ponty's critical character emerges, continuing the Husserlian project of interrogating an indifference toward the presupposed "taken for granted." This is a point that I can connect with Coole's evaluation of Merleau-Ponty's criticality, in that for Coole his bodily phenomenology "exposes the tendency of power to naturalize or render invisible its operations by insinuating them into the taken-for-granted horizons or unquestioned practices of everyday life" (2007, 117).

The interrogation of such presuppositions manifests in Merleau-Ponty's consequent inquiry into the primacy of bodily subjectivity. The aspect of Merleau-Ponty's phenomenology that best informs our considerations here is found in his problematization of subject|object binaries. As early as *Phenomenology of Perception*, it is apparent that for Merleau-Ponty no subject is ever without their body (Merleau-Ponty 1962, 103–4). In *The Visible and the Invisible* Merleau-Ponty applies this insight regarding one's bodily subjectivity to an expansion of the Husserlian position that when one touches one of their hands with their other hand, they simultaneously experience touching and being touched. By interpreting the hand being touched to be both an aspect of the subject, and a part of the world for the subject to touch, the subject for Merleau-Ponty represents both self and world (Merleau-Ponty 1968, 133–34). From this I posit that the embodied subject experiences its subjectivity, as well as the object-world, from inside, and as, the world. Where subject-as-body begins, and the world ends, is perpetually being "renegotiated." This ongoing co-constitution between an object-world and the self informs Merleau-Ponty's consequent definition of subjectivity as "chiasmic" (1968, 130–55). That the subject is part of the world that it touches *as* the touched world is part of the subject means that what manifests is an always internal contact with the world, reconceiving inside-body and outside-world borders. Subjective experience marks a perceptual coherence between body and/as other/thing, "between my body looked at and my body looking, my body touched and my body touching," constituting an "overlapping or encroachment," where "things pass into us as well as we into the things" (123).

From this we can determine that perception that incarnates subjects and objects co-constitutively problematizes the notion of separate, preexisting things that are *then* perceived by a perceiver. Instead, in perceiving, the human body is among such objects/things, where "among" does not simply mean intermingling with, side by side, like marbles jostling for position, but rather suggests that the body is implicated in and through the constitution of other objects/things. Or as Merleau-Ponty eloquently states, "I do not see it [an object] from the depths of nothingness, but from the midst of itself" (Merleau-Ponty 1968, 113). Heidegger's aforementioned *Zollikon* lectures similarly evoke this plastic liveliness of bodily limits. What was implicitly contained in *Being and Time* is here articulated more directly, whereby "when pointing with my finger . . . I [as body] do not end at my fingertips. Where then is the limit of the body? 'Each body is my body.' As such, the proposition [bodily limit] is nonsensical" (Heidegger 2001, 86).

A study, such as ours, concerned with suppositions of secondariness entrenched within the social constructionist time condition, should find a relevance in this impression of the shared timing between a co-constitutive subject and object-world. According to Merleau-Ponty's thesis, perception is not merely what an embodied subject deploys to access already established, predetermined, real objects. Instead, perception marks a world incarnating through its concurrent subjects and objects.

I would argue that another response therefore emerges here to Derrida's critique that phenomenology installs and hierarchizes a logocentric, metaphysical present, seeking a return to the origins of experience. Phenomenological perception is not the discovery of a preexisting origin but a phenomenalizing of origins. Jack Reynolds argues similarly, stating that "what Merleau-Ponty seeks is not merely to return to the phenomenon, but to return to the phenomenon in a way open to that which makes the phenomenon itself possible and . . . that allows phenomenality to be possible at all" (Reynolds 2004, 81). My consequent reading is that if perceiver and perceived manifest co-constitutively and concurrently, then to be present, to see, is to be able to be presented, or to be seen. Embodied perception is an incarnating process that only occurs because it already is of the world, whereby for Merleau-Ponty "he who sees cannot possess the visible unless he is possessed by it, unless he *is* of *it*" (Merleau-Ponty 1968, 134–35; author's original emphasis). In considering the singular, seemingly monistic tendencies being suggested, each materially embodied subject is distinguished from, and as, worldly historical materiality. Merleau-Ponty affirms this kind of monistic reading, describing any perceptual upsurge as a world's perspective that is directed from "self to self, and it traces out an interiority" (Merleau-Ponty 1962, 495). Being materially bodied, in this sense, presents as the internal presenting of a worldly self-perception.

THE MATERIALITY OF THEORY: OBJECT-ORIENTATIONS

What I would like to open now are considerations of the ramifications for our appreciations of theorizing, if I was to correlate this self-perception of/by the world, with theoretical inquiries of/by the world. As with the other contexts of our project, this concerns the social constructionist separation of natural reality from the cultural representations of that reality. Through the influence of Marx's historical dialectic on Merleau-Ponty's outlook, human activity is understood not simply to comprise subsequent reflections on history but rather to manifest materially *as* history. Merleau-Ponty's development of Marx's thesis is that we do not have a history that we can transcend or "end," but that we *are* history. From this I would like to ask whether this critical mode of phenomenological theorization illustrates that human activity, as theory, is a worldly historicized, material self-reflection/self-perception, and what, in later considerations, are the ramifications for climate change discourse if we further conceive of *all human (cultural) activity* (including processes of industrialization) as worldly material activity? As we consider the timing of theory in relation to a material world, so we should reflect on *when* our own theoretical productions and representations as students, scholars, or otherwise, occur, compared to the "real things" about which are theorized. Even for nonacademic readers, the study of the temporal relations between representation and reality can be familiarized in an act such as taking a photograph. Does the image-representation of the world arrive externally after the worldly material reality, or do representation and reality share a more intimate origination?

An impression of theory's worldly historical, material internality seemingly coheres with Horkheimer's original definition of critical theory encountered at the outset of this book, as that which accounts for its own implication in local, and nonlocal, historical production. Furthermore, it seems to fulfill the Frankfurt School requirement that critical theory is not just a scholarly discussion about the world from a position of relative neutrality but rather enacts a *material reconfiguration of that world*. If theory, as I am suggesting, can be viewed as one form of Merleau-Ponty's self-perception *of the world*, then just by manifesting, theory could be that material reconfiguration in which the world perceives itself differently through its enacted self. With this sense of a worldly theorizer, the sequence of a material reality that is followed by the subsequence of theoretical representations of that reality is reconfigurable. It is according to this temporal ramification that we see the relevance of the context of the materiality of theory for an interrogation of the social constructionist time condition in which a natural reality supposedly separately precedes the secondarily culturally produced representations of it.

I am also interested in how current philosophy might view this ramification that theory is not simply an activity of humans that is posterior to the material object reality upon which it reflects but manifests as material reality's self-reflection. This is seen as a continuation of the Merleau-Pontian-Marxist intersection just explored in which phenomenology is portrayed not simply as an existentialist theory but as a critical disruption of what we take for granted about historically material production. Are objects and subjects' theories about objects consubstantially historically materialist?

My intrigue regarding the materiality of theory can be situated in terms of recent philosophical perspectives found in object-oriented ontology. Such perspectives respond to these kinds of queries with concerns about whether reducing the objects with which theory is concerned to a worlding materialization ironically results in neglecting such objects by instead focusing on the process of material, object-becoming. Rebekah Sheldon notes the fear that only attending to a material substratum of *everything* "allows us to skip over objects by seeing through them" (Sheldon 2015, 205). This sentiment is also evoked by Timothy Morton in *Hyperobjects: Philosophy and Ecology after the End of the World* (2013), when he notes that ontological conceptions focusing on materialism and Nature are typically "not fascinated with material objects in all their manifold specificity. It's just stuff" (Morton 2013, 113).

In responding to these reasonable concerns, I would argue, however, that as has been encountered in the preceding Merleau-Pontian discussion, *it is* the "specificity," or the distinguishability, of perceptual objects that conditions subject-and-object-consciousness. Objects are not "just stuff," but importantly particular forms of "stuffing," the Husserlian "this hand" versus "that hand" by which hands are "handed." My consequent point is that theory, as perception, represents one such genre of specific objects, manifesting simultaneously as a distinguishable perceiver/touching-hand/theorizer and perceived/touched-hand/theorized. This is a point that object-oriented ontology will endorse in positing that philosophical theory is not restricted to a theoretical or representational construction about a prior, separate, material reality but rather represents a broader material*izing* practice. As Ian Bogost suggests in his essay "The Aesthetics of Philosophical Carpentry" (2015), philosophy is not restricted to "words written down on paper, typeset, and glued to binding" (Bogost 2015, 83). In rethinking the object boundaries of philosophy's or theory's identity, both in the way philosophy materializes as well as of how entities other than humans participate in this philosophical materialization, a dispersed sense of philosophizing is endorsed. Bogost's discussion of nonhuman philosophers and theories accords with my reading of the Merleau-Pontian structure in which philosophical theory represents a world's rather than merely a human's self-reflection.

We see a similar kind of inquiry regarding object-oriented ontology's focus on the philosophizing object in Graham Harman's *Tool-Being: Heidegger and the Metaphysics of Objects* (2002). A crucial feature of Harman's motivation to examine Merleau-Ponty's phenomenological predecessor concerns an attention on how Heidegger "reminds us that theory is not to be attained by a mere gazing at present-at-hand objects" (Harman 2002, 57). Harman instead recognizes in Heideggerian ontology a process beyond chronologically linear, sequential restrictions of the human discovery, reception, and theorization of docile objects. Instead what is acknowledged is the way in which worldly objects, and theories about those objects, grasp and are grasped together. Theory and world coincide, or in Harman's terms, "theory emerges out of being-in-the-world" (2002, 57), whereby the object-experience of being-things is a theoretical reality, not merely a representation of reality.

This object-substantiality or materiality of theoretical methodology is equally posited by thinkers associated with "new materialism." Karen Barad's examinations of the entanglement between matter and meaning in *Meeting the Universe Halfway* (2007) are notable in this regard. The argument that the theorization or observation of worldly objects is not simply a case of a perceiving, culturalized subject, overcoming a representational distance or separation *between* themselves and the natural or real object to be "known," emerges in Barad's interpretation of the conditions of quantum mechanics. In reviewing how physics examines the wave or particle status of atoms of light, Barad duly adopts the interpretation that the arrangement of the experiment does not passively observe but participates in the production of the wave or particle behavior being investigated. This inseparability of observing and observed components allows Barad to present an argument of *intra*action between co-implicated worldly entities. Barad extensively substantiates her position according to its coherence with Niels Bohr's assertion that "in the face of quantum nonseparability," "real knowledge" of worldly objectivity is not based "on an inherent or Cartesian cut between observer and observed" (Barad 2007, 339). The involvement of observing and observed entities in and as each other means for Barad that the apparent borders of an entity, rather than marking an exteriority from the entities that are examining it or theorizing about it, instead manifest *with* such examination/theorization, whereby "'observer and observed' are nothing more than two physical systems intra-acting" as the one system (2007, 339–40).

In exploring how discursive practices *produce* rather than merely *describe*, Barad presents work in all areas of the humanities and the sciences as examples of "material (re)configurings of the world through which the determination of boundaries, properties, and meanings is differently enacted" (Barad 2007, 151). Barad conjures Merleau-Ponty's sense of a worldly self-inquirer to this end, describing epistemological endeavors, or "knowing," as "a matter of part of the world making itself intelligible to another part" (2007, 185). Just as I have argued that for Merleau-Ponty perception is only possible because of an implicated co-production between subject and object in which each constitutes the other, equally for Barad we only know the world "because we are of the world. We are part of the world in its differential becoming" (185). In representing her voice as coming from a philosophical and sociological standpoint, such fields are argued to be capable of reflecting upon their methods and preconditions. Barad connects this to a disciplinary meta-analysis also identifiable in the sciences, noting that physicist Bohr's recognition of how one's physicality affects experimental considerations positions any consequent theorizing as an "embodied practice" rather than as "a spectator sport of matching linguistic representations to preexisting things" (54).

These considerations reveal how philosophy can acknowledge a material, theoretical reciprocity between inquirer and inquired in which a subject's relation to the world is not conditioned by the inaccessibility of approximated social constructions of what is separately, actually, materially real. Unsurprisingly, this perspective is not adopted by all philosophers without caution. Impressions of a simultaneity between an inquiring subject and an inquired object leads Dorothea Olkowski, for instance, to pose an important question regarding the concurrence of these temporal directionalities. This concerns what is implied in both the act of seeing and in the recognition of being seen and queries whether a subject can perceive the world *and* apprehend a perception of themselves *simultaneously*? (Olkowski 2010, 532). What does it mean to be a subject, to see, if to be a subject is always, to some degree, to be implicated in, and as, the object that is the focus of the seeing subject? Olkowski notes that it could be presumed that only one of these actions, either seeing or acknowledging being seen, is possible at one time. If to perceive is to consciously "look at or to see something in a deliberate manner . . . and if to apprehend the look is to be looked at and to become conscious of being looked at" (2010, 532), then there seems to be a conflict, rather than a simultaneity, between such states. We need to venture into Merleau-Pontian notions of ambiguity, and indeed the ramifications for climate change discourse of this ambiguity, before I am able to respond to this valid concern from Olkowski.

Through a phenomenological argument I have considered how the body, constituted by worldly material historicity, chiasmically perceives itself (from) within this worldly material historicity. The body reveals Being's *ambiguity*, at once distinguishing both the subjectivity of self-as-material-history and the objectivity of world-as-material-history. Such "reversible" worldly perception is *conditioned* by what Merleau-Ponty calls *the flesh of the world*. This generalized flesh of perceptual becoming is not materially present, instead conditioning the world perceiving itself by constituting what is simultaneously common and yet distinguishable about worldly entities. As Gail Weiss's chapter (Weiss 2009) on ambiguity explains, the body-as-world-as-flesh is at once particular *and* general, local *and* global, or "subjective and at the same time, pervaded by what Merleau-Ponty calls an 'atmosphere of generality' that connects one to all other bodies, human and non-human" (2009, 134). This notion of "generality" seemingly refers to the explanation from Merleau-Ponty in *The Visible and the Invisible* that the flesh of the world, rather than being restricted to a specific material/matter, is the primordial "element" that "stylizes" all materiality. As he states,

> The flesh is not matter, is not mind, is not substance. To designate it, we should need the old term "element" . . . in the sense of a *general thing* . . . a sort of incarnate principle that brings a style of being wherever there is a fragment of Being. The flesh is in this sense an "element" of Being . . . the inauguration of the *where* and the *when*. (Merleau-Ponty 1968, 139–40; author's original emphasis)

As we see in Merleau-Ponty's passage above, it is Being's self-stylization which inaugurates all "wheres" and "whens." This fleshed stylization is therefore Being's differentiation, consistent with my assertion to come that worldly distinguishability conditions (inaugurates) there being Being. As I read this phenomenologically, worldly flesh is not reducible to particular materialities/spatialities. Rather, the ambiguity of worldly flesh, a self-reflective, theoretical, practical becoming is the inaugurating reflex through which monistic materiality manifests.

Via Merleau-Pontian phenomenology we can appreciate that when entities, such as (but not exclusively) embodied humans, manifest, they do so as the perception of the world, a perception that concurrently produces the world. The human, and the human perception, marks how perception is fleshed, production is fleshed, where flesh facilitates a world distinguishing itself from itself. This is why I argue that the critical theorization of Being, the critical perception of the human-as-perception of Being, is not that which arrives straightforwardly secondarily to, or even late to, an established worldly/real scene into which it intervenes and reconfigures. Rather, theoretical perceptions represent one of the fleshy conditions by which Being is always already self-perceptually becoming/critiquing Being.

I wish to now test the rigor of these considerations by applying them to the context of climate change discourse. What does an argument like this mean for how we perceive, and theories about, a changing environment? Is such theory, not to mention the climatic objects of environmental change that theory targets, also indicative of Being's self-perceptual *becoming*? This is seemingly a precarious proposition, given that as we are about to explore, these theories often suggest we are conversely, potentially, "running out of time" to address such climatic shifts before we might no longer be being/existing at all. With the Marxist notion of the end of history in mind, it must be asked whether climate change discourse suggests that there is a dangerously impending, literal end of human history.

CRISIS CRITIQUES: CLIMATE CHANGE REPRESENTS A PLANET "RUNNING OUT OF TIME"

The industrialized Earth's seemingly adverse climatic changes are often reported with the accompanying alarm that the planet is "running out of time." What is posited to be running out of time in such reports is not simply the planet itself. Instead, the existence of humans is either directly or indirectly said to be jeopardized in that as the Earth's climate changes, so the planet might become humanly uninhabitable. From this manifests the sense of a conflict between the ecological and climatic transitions of the Earth and human life. Complementarily, time is presented as an oppositional force against which humans are racing.

As has been noted in earlier chapters, the characterization of time as both a transcendent source and as something against which we battle or race permeates our everyday experiences. It governs short-term intentions such as doing enough work in a day or arriving promptly at social commitments, as well as longer-term ambitions that we foresee for our lives. What will be considered now though is whether climate change discussions that demand time is something of which humans are "running out," indicate that humans are involved in an adversarial relation with time at an existential level. Is lateness not simply a socially constructed, relative marker about daily responsibilities, but also a reflection of our tardiness in saving a planet that is running out of time, our time, entirely?

The argument that Earth is "running out of time" has had no more prominent champion than Denis Hayes, the organizer of *Earth Day*, the "event which gave birth to the environmental movement in modern America" (Quade 1990, 16). Hayes, described by Quade as North America's most prominent environmentalist, is joined by Donald Brown, associate professor of environmental ethics, science, and law at Pennsylvania State University, who similarly proclaims, in response to the *Copenhagen Accord*,[7] that scientific evidence concerning climate change reveals that the "world is running out of time" (Brown 2010). Nick Hansen, introduced as the world's leading climate scientist in *Storms of My Grandchildren* (2009), observes that contrary to efforts to regulate the human contribution to ecological/climate change, "short-term special interests" neglect that "we are running out of time" (Hansen 2009, xi). This characterization of ecological/climate change is not a recent trend. Over forty years ago, Albert Engel, prominent geoscientist and author of "Time and the Earth" (1969), warns that "man, if not the Earth, is running out of time" (Engel 1969, 460). Furthermore, when David Runnalls, as president of the *International Institute for Sustainable Development*, declares that "the world is running out of time to deal with seemingly overwhelming environmental threats" (Runnalls 2008, 19), it is made apparent that human existence is included in the phrase "the world."

This vernacular suggesting that the human response could be too late to address the existential threats posed by climate change is also evident in work found in the environmental humanities. Jonas Anshelm and Anders Hansson's "The Last Chance to Save the Planet" (2014), published in the pivotal journal *Environmental Humanities*, notes that in discussions concerned with responses to climate change and "the gravity of the situation . . . it is claimed that it might already be too late" (Anshelm and Hansson 2014, 107). For Deborah Bird Rose, this notion that "existence is running out of time" is embedded within the scholarly navigations of the environmental humanities. While Rose notes that the "environmental humanities is a rapidly growing and evolving project," it is always invested with a sense of "urgency," given that "time is running out for sustainable life on earth" (Rose 2016). It must also be recognized, however, that the environmental humanities present a multifaceted approach to the advent of climate change. Eben Kirksey's *Emergent Ecologies* (2015) rejects the apocalyptic tone often given to discussions regarding environmental transition, instead identifying the opportunistic ways in which new ecological relations manifest as a result of climate change. My work in this chapter will not be focusing on finding literal examples of ecological prosperity as Kirksey does. Nevertheless, I will be directing a discussion that interrogates, rather than advocates, the separatist, temporalizing conditions that I believe lurk within fearful appraisals of current human-ecological relations. As with the contexts of the cultural relativization of time and of body modification, by engaging climate change discussion we are concerned with the tones of secondariness by which the temporality of human activity is defined against a natural backdrop.

The constructionist position that we are interrogating, that the source of time is an inaccessible, adversarial externality, grounds presumptions that we can only race against it. Barbara Adam's "Time and Environmental Crisis," in engaging the "consensus that we are facing an environmental crisis," notes that "while the spatial dimension has been brought to the fore in a number of disciplines, the temporal equivalent has stayed implicit" (Adam 1993, 399). My focus on the relation of time to environmental change is in this regard novel. Another of Adam's inquiries, while differing in ways that will become apparent, also recognizes that an "explicit focus on time . . . illuminates the shadow side of environmental phenomena, aspects which are normally ignored" (Adam 1994, 110). I believe Adam here touches on the belief that we can affect the materially spatial realm, whereas time is objectively separate, harder to locate, and beyond human influence. This evokes the earlier discussed descriptions of time as invisible, mysterious, and hidden.[8] The impression to which I point here is that what is material and spatial about the world, that being its ecology, can be both adversely affected and beneficially repaired by human material resources. Time, on the other hand, can either be utilized or forgone; it functions beyond the realm of material intervention, whereby the only option for humans is to race against it. Adam evokes this conviction, or rather this fear, of humans, that "we are running out—not of resources, but of time" (Adam 1993, 401). Humanity is characterized, in its untimely governmental intervention and consumer behaviors, to be collectively late to address what matters about environmental survival. That the entire species could soon be "late" in the mortal regard due to environmental change is the implication.

Here we are reminded of the earlier observation of a second relevance to the "late" theme that we are engaging. Complementing the primary investigation into the secondariness by which the lateness of social phenomena is invested, the signification of lateness as death/mortality has emerged in discourses concerned with natural origins, with body modification practices, and now, with the effects of climate change. Humans are characterized as late *in the present* in terms of their collective response to climate change, as well as potentially late (dead) *in the future* if such environmental damage jeopardizes their existence on Earth.

This jeopardy is routinely portrayed via the correlation of changes in the environment with an uncertain future for the human, and other, species. Increasingly polluted air, escalating heat, melting polar ice, and rising sea levels are all positioned as indicators of a transitioning ecology that is incompatible with the possible life of subsequent human generations. While these physical changes reference the overt forms by which climate change is said to manifest, Merleau-Ponty's phenomenology can help us to intervene where a blindspot might preside regarding what is taken for granted about the fundamental constitution of the objects of climate change. With Merleau-Ponty, we have characterized humans as constituted by a worldly fleshing: the flesh of the world. By what, however, is climate change constituted?

Raymond Pierrehumbert, lead author on the third assessment report of the *Intergovernmental Panel on Climate Change*, explains in "Climate Change: A Catastrophe in Slow Motion" (2006), that "human-induced" changes to Earth's ecology and climate are attributable to modern industrialization. The most significant change has occurred via global warming, "wrought by industrial carbon dioxide emissions" (Pierrehumbert 2006, 573). While "several other gases" contribute to the "human-induced" aspects of global warming, carbon dioxide is "by far the biggest player" (2006, 274). Albert Engel likewise commentates as far back as 1969 on the destructive influence of industrial, carbon emissions pouring "millions of tons of carbon dioxide into the air from the irresponsible destruction of fossil fuels" (Engel 1969, 480–81).

Carbon dioxide is an atmospheric constituent that has played a repeatedly prominent role in the Earth's climatic changes. This includes the Ice Ages, the warmer dinosaur era 70 million years ago, and the Earth's collapse into deep freeze in the Neoproterozoic era 600 million years ago (Pierrehumbert 2006, 574). However, human industrialization, we are told, has increased the atmospheric carbon gas levels by about fifteen times their preindustrial levels (2006, 574). Global news in recent years has been dominated by this theme. *The Atlantic* magazine describes carbon dioxide "as the great engine of climate change" and that the dangerous level it reached in November 2015 means that "Earth's atmosphere just crossed an epochal threshold" (Meyer 2015). The United Kingdom's *Guardian* newspaper reports that the amount of carbon dioxide gas in the Earth's atmosphere is more than it has experienced in 5 million years (Carrington 2013). Consistent with how climate change discourses characterize the world as "running out of time," Australia's *Sydney Morning Herald* describes the same story as a worldly "clock running down" (Hannam 2013).

A consensus thus emerges identifying carbon emissions both as the enemy to the Earth, and that which humans must combat. Instead of contributing yet another voice to this chorus, however, what I am intrigued by is what occurs when we recognize that carbon emissions, an inherent part of climate composition, are also a fleshy manifestation of Being. The stuff of carbon dioxide is a stuff of the world. Carbon dioxide is constituted by that which is *of* the world. I am compelled to raise this possibly rudimentary point, given that in terms of the phenomenological method through which we have appraised the world's becoming a world, every materializing or "stuffing" form, in distinguishing Being/world, is the worlding that is internal to, and *is*, Being/world. I am conscious of the earlier insight that Merleau-Ponty is careful to not entirely reduce the perceptual, reversible flesh*ing*-of-the-world, to materiality. When I describe such materialities accordingly, I am not restricting the word "flesh" to an unrefined, literal definition of corporeally specific ontic bodies. Rather, any material body is a form or "style" of this flesh, not as an impossibly purely present entity but as that which ambiguously materializes, fleshes, or bodies subjects and/as objects. In now describing the "materialities of climate change," what are at stake are thus not simply the tactile forms in which climate change *presents* but rather the presenting-as-fleshing by which climate change materialities occasion a world. Climate change's presenting-as-fleshings, its tactilities, manifest as certain "styles" that Merleau-Ponty describes of flesh.

Our earlier review of discourses attending to climate change has revealed the conception of a combative opposition between humans and climate change. The increased emission of climate change materialities, such as carbon dioxide, is inversely matched to the decreased amount of time that humans and the Earth are described as having remaining. However, if climate change materialities are forms of worldly materializing/fleshing, and like all other materializing/fleshings they reversibly, perceptually co-condition the being of Being, then interpretations of an adversarial incoherence between materially embodied humans and materially climatic conditions exhibit a certain vulnerability. It is in the materializing co-conditioning earlier evidenced phenomenologically that we might suspect that the materialities of climate change have a more generative relation to our time than simply symbolizing that we, or our planet, are running out of it. If this sense of existential temporality is problematized, then the argument that socially or politically motivated responses to such a crisis are occurring too late will also invite reconfiguration.

My interest in Being's self-distinguishable, corporeal *ambiguity* concerns how it informs the commonality rather than the representational inaccessibility implicit to the human experience with the world, and how this in turn attends to the question of natural-social temporal relations. This is inspired by Merleau-Ponty's summation that "the ambiguity of being-in-the-world is translated by that of the body, and this understood through that of time" (Merleau-Ponty 1962, 98). Instead of a look *by* a human-as-body *at* a separately originated world, Merleau-Ponty asks us to consider how the subject-body and the object-world temporalize with, as, and through each other via Being's look. My argument is that this look, in a form such as a human-as-body, or the situated perspective of a theory, perceptually distinguishes the world from itself. Indeed, it distinguishes Being from Being, a concurrent separation-and-implication, which by visibilizing particular thing from/as everything, is what Merleau-Ponty describes as "constitutive for the thing of its visibility as for the seer of his corporeity" (Merleau-Ponty 1968, 135). Given this ambiguity, the corporeal human body, or the theoretical body of work, chiasmically perceives the world from within a worldly materializing from which it is distinguished and which it concurrently distinguishes. Theory's perception of its embedded history, of the conditions of its germination, a parameter already flagged as central to the characterization of theory as critical, is here apparent. This perception is not simply *of* what conditioned the materialization of a self, or of a theory, but is a perception which *is* those conditions. To see Being/world, as theory does, is to see theory's historicism, the "by which" that reversibly means theory is able to be seen. In materializing, theory makes visible, it materializes/visibilizes, a world.

My argument here is that Being, in self-distinguishing, becomes. Without this perceptual self-distinguishability, Being would not become by emerging from itself as the self it always already was. I argue that Being-as-self becomes what it "always already was" given that its upsurge comes from nothing but itself. This "always already was" does not replicate a past version of itself, however, given that as we have seen in earlier chapters, the past is always being (re-)produced, whereby a past "in itself" would contrarily not be a temporal state. This offers one response to Olkowski's earlier insightful commentary concerning the curious, simultaneous temporality of corporeal seer and seen. If time's source externally, sequentially, preceded the incarnation of subjects, such subjects would be bound to the jurisdiction of an exclusively forward-moving temporality at any particular point of which it might only be possible to either consciously perceive or be conscious of being perceived. The unidirectional progression of time would mean that such points would be mutually exclusive. However, when the source of time *is* the world's self-divergent, self-productive, self-perception, seeing and being seen can only be simultaneous and co-constitutive. The world is incarnated by perceiving itself through its worldly perceiving entities. If this conditions the being of Being, Being's origin, must this be Being's time?

My reading of Merleau-Ponty is that time's origin is not the disruption of Being's atemporal *prior* presence. Instead, wherever there is Being, wherever there is any *thing* or *thinging*, any stuff or stuffing, there is time. Being's wherever is its whenever. In terms of humans-as-things, time manifests not as a passage acknowledged by an onlooking subject or theorist. Rather, time, as embodied subjectivities, is something Merleau-Ponty acknowledges as early as *Phenomenology of Perception* as "nothing but a general flight out of itself" (Merleau-Ponty 1962, 487).

The ramification is that every subject, human or "otherwise," is inescapably, but not submissively, temporal and temporalizing by perpetually distinguishing what would otherwise be the atemporal realm of an "unbroken chain of the fields of presence" (Merleau-Ponty 1962, 491). This coheres with the argument developed in earlier chapters, which illustrated the unavoidable participation of human subjects in the temporality of objective rhythms. Speaking to the earlier caution against interpreting time's emergence as the disruption of a prior plenitude, Merleau-Ponty's use of the term "unbroken" should not be read as an anterior plenitude that is dismantled by time as a *subsequent* event. Rather, time is the distinguishing break that was always already occurring in order for there to be Being and/as time. The temporal "break" cannot disrupt a prior plenitude, given that without the break, without time, there is no plenitude, there is no Being.

The phenomenological intervention I have directed us through to this point characterizes human subjects, and all other entities, as inescapably temporal, but not because of an unavoidable finitude or transience that a transcendent, "inaccessible," time-source imposes. Rather, all materializings/fleshings are inescapably temporal because *they are temporalizing*, whereby from their intersubjective, co-constitutive incarnation, time and/as Being manifests. In terms of climate change discourse considerations, it could be argued that a changing ecology does not therefore indicate that time is running out for the planet. Where Nature manifests, where there is materializing, even in the form of humanly uninhabitable environments, time manifests. In considering my earlier appraisals via new materialisms and object-oriented ontologies of the materiality of theory, it could furthermore be argued that where this materiality manifests, even without a human author, theory as perception, as a worldly reflection, also manifests. The most critical of recognitions could in this regard be that the advent of the late/dead human species would not spell the end of theory.

In chapter 1 we reviewed impressions of the contingency of social times and associated structures regarding lateness, whereby all such social times were positioned as subsequent manifestations of an anterior, naturally worldly, phenomenon. From this came the insight that the social time structures that lacked strict protocols regarding lateness were conceptually restricted to being closer to natural time. Conversely, their tardiness-punishing, global counterparts were viewed to have advanced beyond this state. In this chapter though, we see via phenomenologies and associated philosophical perspectives that an intervention can be made regarding the hierarchization of social time structures according to the supposedly greater alignment of any "social time" with natural time. Thanks to the platform provided by Merleau-Ponty, I posit that it can now be argued that if all materialities, all entities, flesh the time-of-the-world, then all human societies manifest with, and as, time's naturally material conditions. This indicates a more egalitarian appreciation of social times globally and evidences further grounds perhaps for the critical portrayal of Merleau-Ponty's phenomenology.

A potential contradiction has emerged, however. If each individuation of Being is the incarnation of time, are localized or subjective temporalizations required that cannot be accommodated within the objectively common features of the time underpinning social synchronization? When every thing is a time or a timing, how do plural things co-constitute a shared time? In exploring this question of synchronization, the social constructionist time condition that we are examining in this work, which characterizes social time as contingent, and separated from what is naturally real about time, will be unsettled through Merleau-Ponty's considerations of the consciousness of time. This is because, if as has just been phenomenologically explained, the material ontology of time has intersubjective and interobjective conditions, then there seems to be something inherently social about such time.

Given the already developed position that Being's self-incarnation is constituted by a perpetual self-perception, I would now assert that what is also constituted is Being's self-awareness/self-consciousness. I cohere on this point with Merleau-Ponty's appreciation of consciousness as a self-self production, where the world in its fleshing of self and/as selves becomes "aware of itself, for the explosion or dehiscence of the present . . . is the archetype of the relationship of self to self" (Merleau-Ponty 1962, 495). Self-perception and self-consciousness are reduced to a singular mechanism immanent to Being's reversible, perpetual self-touch. Could we equally say, therefore, that theory, when reflecting on its own material and historical conditions, is also a form of worldly awareness? From where does theory manifest, after all, but from perceptions of the world, as the world, a point that might remind us of Merleau-Ponty's contestation to Descartes's dream skepticism.

We have discussed how this self-incarnation of Being produces time. If, as is apparent in the preceding determinations, Being's self-incarnation is Being's self-consciousness, does this mean furthermore that each Being-incarnating, or Being-temporalizing, subject is never outside time-consciousness? That flesh (Being) is aware and conscious of itself would indicate that just as in Augustine's, Durkheim's, and Halbwachs's, models, consciousness for Merleau-Ponty is not a brain-demarcated cognition. Merleau-Ponty presents a consciousness that overlaps with(in) itself as worldly perception. This might clarify his characterization of being-in-the-world as "unreflective," a term about which I was earlier concerned. Being-in-the-world's perceptive ontology is not unreflective in opposition to ration/cognition/consciousness. Instead, Merleau-Ponty employs "unreflective" to distinguish being-in-the-world's self-incarnating-fleshy-thought (where thought *is* worldly incarnation), from conceptions of a reflection that is restricted to a subject's thoughts of a *separately incarnated*, fleshy world. Merleau-Ponty's phenomenological consciousness can be used here, I believe, as a critique of the social constructionist position being interrogated in this book, in which a preceding, inaccessible, real arena is misrepresented by culturally contingent cognitions and perceptions of it. Conversely, for Merleau-Ponty perception *is* real ration/cognition/consciousness, as "one must see or feel in some way in order to think . . . every thought known to us occurs to a flesh" (Merleau-Ponty 1968, 146).

Such an all-encompassing model disperses the social, collective consciousnesses posited by Halbwachs, now through a physical dimension. Nick Crossley shares my opinion, whereby in discussing a carnal sociology, Crossley states that Merleau-Ponty allows sociological practitioners to "understand that human agents-subjects are bodies, and that bodies are sensible-sentient, communicative, practical and intelligent" (Crossley 1995, 60). Bodies-as-subjects are sentient of time I duly argue, because they are sensible *as* time. Time is perception, a materializing, manifesting as a general worlding that it constitutes and by which it is constituted. Merleau-Ponty puts this in the terms that "time must *constitute itself*—be always seen from the point of view of someone who *is* of *it*" (Merleau-Ponty 1968, 184; author's original emphasis). Indeed, in Merleau-Ponty's impression, "consciousness" has no meaning independent of this ontological self-relation that is always "affected by itself or given to itself" (1962, 495–96).

The consequent "corporeal ambiguity" upon which Merleau-Ponty's phenomenology relies can now be embraced as a perpetual immanence that facilitates, rather than jeopardizes or inhibits, knowledge of self and/as time. Ambiguity represents one's self-relational becoming, whereby for Merleau-Ponty, I "know myself only in my inherence in time and in the world, that is, I know myself only in ambiguity" (Merleau-Ponty 1962, 402). That subjectivity here is conditioned by the *divergent* ambiguity of a self-world co-constitution contradicts Derrida's criticism that a metaphysically hierarchized, logocentric, binary present underpins all phenomenology. Jack Reynolds agrees, asserting that if subjectivity is predicated on "a difference that is not a dualism . . . Merleau-Ponty does not appear to conform to the standard deconstructive definition of the metaphysics of presence" (Reynolds 2004, 58).

Could this sense that consciousness is, and has, a *worldly* resonance, destabilize the assumption that humans are the only material incarnations with a self-awareness of their role in, and relation to, ecological/climatic change? Here we must return to work of Pierrehumbert, who characterizes humans as an environmental force, causing "practically irreversible changes in global conditions" (Pierrehumbert 2006, 573). This is not to say that humans are the only worldly incarnations to cause globally significant ecological/climatic change though. The evolution of oxygen-generating photosynthetic algae between 1 billion and 2.5 billion years ago "changed one fifth of the atmosphere, poisoned much of the previous ecosystem, and terminated the dominant role of methane as a greenhouse gas" (2006, 573). The colonization of land plants half a billion years ago had similarly momentous repercussions by increasing "the rate at which atmospheric carbon dioxide is converted to limestone in the soil, leading to severe global cooling" (573). Nevertheless, despite how different worldly incarnations, different forms of life, have changed ecologies and climates, Pierrehumbert separates "human induced" changes from all others. What makes current climatic change unique, according to Pierrehumbert's thesis, is that "the causative agents—humans—are sentient" (573).

The question of whether plants and other nonhuman entities are sentient has in recent theory been prominently engaged by object-oriented ontology.[9] If we follow the dictionary definition of sentience as "being conscious of sense impressions" (Merriam-Webster 2013), it could in fact be concluded that plants and algae, conscious of their sense impressions, meet necessary criteria of sentience.[10] Rather than becoming too involved in this debate though, what we need to flag is how doubts about sentience beyond humans perpetuate the assumptions implicit to the social constructionist time condition, in which humans are primarily, and perhaps exclusively, positioned as being capable of representing reality. The human consciousness of such reality, apparent in human culture's representations, is interpreted to be *uniquely divorced* from the reality of the objective, natural world, accordingly.

In considering phenomenological meta-reflections in this chapter, it would be disingenuous when discussing consciousness to ignore the social phenomenology of Alfred Schütz. This is because, consistent with above impressions of the separate status of the human consciousness, Schütz posits a "unique stream of consciousness of each individual" that is "essentially inaccessible to every other individual" (Schütz 1967, 99). Schütz's phenomenological consciousness is informed by Husserl's notion of intentionality (in which a belief must be about, or directed to, "something"). The intersubjectivity of intentionality is in this understanding a transcendent relation. Intentional experiences are of "externalities" to one's consciousness, such as "intentive mental processes belonging to other Egos with other streams of mental processes, and likewise all acts directed toward physical things" (Husserl 1983, 79). The social constructionist time condition's installation of an inaccessible, naturally material arena-of-things is duly apparent.

Schütz (as with Husserl) postulates external conditions when discussing intersubjective perception. Indeed, Schütz takes a citation from Husserl's *Logical Investigations: Volume 2* (2001 [1929]) to exemplify this:

> The listener notices that the speaker is expressing certain subjective experiences of his and in that sense may be said to notice *them*; but he himself does not live through these experiences—his perception is "external" rather than "internal." (Husserl, quoted in Schütz 1967, 100; author's original emphasis)

It is this belief in an in-itself externality from human consciousness that Merleau-Ponty's reconfiguration of the Husserlian project disrupts. Perceptions in the Merleau-Pontian sense are of something less polarized than a social constructionist's distantly external bodies, which Schütz believes one only "notices" as "merely images" (Schütz 1967, 102). Accordingly, I argue that Merleau-Ponty's phenomenology facilitates an interrogation of a supposed opposition between humans, and climate change materialities, where the proliferation of the latter is characterized as a result of humans responding "too late" to an environmental and existential threat.

ENVIRONMENTALLY SYNCHRONOUS, MONISTIC, SOCIAL CONSTRUCTIONISM

An example of an oppositional characterization of the human-climate relation is found in a key claim made by Barbara Adam. Adam perpetuates the concern "pervading climate change debates" of "time running out for effective action" (Adam 1993, 401) by arguing that organized human (often political) responses to human-induced climatic transitions are too slow. By failing to recognize the "exigency of the crisis" (1993, 401), what results for Adam is a world of "out of sync" relations. Having not stopped the adverse effects of the human presence on the natural environment, humans are defined as existing asynchronously with the changing planet (401).

In "Running Out of Time," Adam further discusses how all human-induced environmental, ecological, and climatic, changes, are characterized by dissonant time-frames (Adam 1994, 98). Nowhere is this made more apparent, "Time and Environmental Crisis" informs us, than in the "depletion of the ozone" (Adam 1993, 401). Ozone depletion is significantly increased by chlorofluorocarbons (CFCs), synthetic gases found in appliances in most Western households (1993, 401). Consequently, the speedy removal of all CFCs from all markets would significantly assist in arresting ozone depletion. That this has not occurred exemplifies, for Adam, the aforementioned "out of sync time-lags" (402), marking the lateness of the human response to our own existential crisis.

Considering preceding phenomenological considerations, I propose the reading that CFCs, as manifestations of worldly materializing/fleshing, are internal to the world's self-distinguishing ontology. This is an ontology in which materializing/fleshing humans are equally present. Adam forwards an awkward logic in which if our "contemporary environmental crisis exhibits global features," then they must be the duality of "ecologically networked interconnectedness" *and* "out of sync time-frames" (Adam 1993, 401). From the phenomenological perspective with which we are engaged, though, we might express concern about how an "ecologically networked interconnectedness" could ever exclude or be out of sync with materializations such as CFCs. If the humans who make CFCs and the materials which comprise CFCs all manifest as a worldly flesh/fleshing from which subjects and objects emerge concurrently, where is the asynchronous "outside" of worldly origination for any such manifestations? This chapter has attempted to problematize such inside/outside modes of analysis, stemming from the underlying inquiry into the constructionist condition that time's source is an inaccessible object-reality that is outside the social and subjective (mis)representations of it. Adam in fact is critical of the domination that the sequence which is integral to the socially "linear-perspective and clock time" has, describing both as "powerful externalizers that separate subject from object" (1993, 409–10). Despite this, I assert that Adam's thesis of out-of-sync temporal relations *does* externalize subjects from objects, humans from CFCS, humans from Nature, and misrepresented time from actual time. Where, for Adam, humans and their cultural practices arrive late to, and disturb, a naturally occurring scene with which they are not synchronized, so pervade presumptions of a linear separation.

We are instead invested in a study of whether the corporeally themed adaptation of the phenomenological method facilitates the recognition that all incarnations of the world are co-constitutively in sync. Depending on how ecologically and materially dispersed is one's sense of a community, the usual definition of social synchronization can be destabilized from this imperative. This usual definition, I argue, is consistent with the social constructionist time condition's separation of real time from social time(s). I say this because in societies with a standardized or centralized time, social synchronization is typically conceived in terms of its representational conditions on clocks and calendars. Subjects meet by using the same, socially derived representation/construction of a natural, celestial rhythm. When time is only understood as divisions along a linear path, as separate minutes, hours, days, and years, each subject only synchronizes with another subject if both know the same representation of a time from whose source they are separated (guaranteeing its objectivity for both subjects). Such conditions also facilitate adjudications on social synchronization's contrary event: lateness. In Adam's view, this is a restricted interpretation of time that needs to be superseded, in that "we need to overcome the clock-maker's reductionist view of nature and society" (Adam 1993, 411). This directive appeals to my interests. Less similar to my position is Adam's assertion that "time-lags" are shared by "out of sync" entities. The implied temporal abyss, an *insurmountable asynchrony*, requires in my contention an externally antecedent, natural time-source, which humans can only ever distantly estimate and misrepresent, and that their materials disrupt.

By instead engaging Merleau-Ponty's phenomenology of time, I want to open a characterization of synchronization as a naturally immanent co-incarnation, rather than as what is conditioned by a representational time-structure that itself is conditioned by a separately natural rhythm. My argument is that materializing subjects and objects are synchronous because their "arrival" is characterized by how they originate concurrently with, and "co-constitutively/intersubjectively as, each other (socially) as time. What the ramifications are for the fact that, despite this phenomenon that I will call an "originary synchrony," social lateness still has a daily reality in our lives, will be explored at length in the final chapter. What is recognizable now, though, is that all materializing is perceptually, synchronously originating. Our frame of reference regarding synchronization is not circumscribed only by a separate symbolic gauge found in human constructs but is heralded by the concurrent co-incarnation of *all* timings. The materialities of climate change, in systemically plurally co-constituting other materialities (as time[s]), duly manifest in sync with, rather than adversarially opposed to, humans.

Could this monistic sense of synchronization therefore be described as a "*social* synchronization"? In revisiting an earlier discussant, George Herbert Mead broadens his definition of socialization-via-pluralization in a way that we might believe affirms that the just reviewed synchrony *is socially* constituted. Mead removes certain limits from conceptions of sociality in stating that sociality "belongs not only to . . . human organisms" (Mead 2002, 46). What Mead approaches on this theme is a belief in an unrestricted social arena, in which every entity, ecological or otherwise, could be considered a constituent (2002, 177). The social emergence of embodied organisms and/as "physical things other than the body" (136), incarnates all contemporaneously. The human utilization/implementation of supposedly inanimate physical/material objects such as tools, for instance, emerges as a pluralistic (socialistic) emergence of humans *with* such objects. Mead's conception of an unrestricted social realm duly contradicts accounts of a human>object hierarchy as advanced by Sigmund Freud (1989 [1930], 42). Freud posits that "the first acts of civilization were the use of tools" (1989, 42), hierarchizing humans over other animals and inanimate objects (42–46). Conversely, Mead sees no such hierarchy, instead positing an unrestricted social emergence. This systemically plural, all-encompassing co-social-constitution means for Mead that "the bodily selves of members of the social group are as clearly implemental as the implements are social. Social beings are things as definitely as physical things are social" (Mead 2002, 177).

While this dispersal of sociality is galvanized by object-oriented ontology's earlier exhibited, critical awareness, to not confine agency to humans, there are associated interpretations in which this position is only conditionally approached. Bruno Latour posits a hybrid, social ensemble, of human *and* nonhuman components, a network of (re-)distributed agency comprising a "new assembly" (Latour 1993, 145). While Latour accommodates "nonhumans" within this social fabric, where "the human is not a constitutional pole to be opposed to that of the nonhuman" (1993, 137), the division of preexisting human and nonhuman elements that are *then co-joined*, endures. This is perhaps most explicit in Latour's criticisms of the dangers that deconstruction and sociobiology pose in their affirmation of "what has not been built at all by any human hand" (Latour 2003, 41). Vicki Kirby consequently describes Latour's "hybridity as a composite of *both* 'human' *and* 'nonhuman'" (Kirby 2011, 85–86; author's original emphasis), and critiques the implicit human>Nature hierarchy of its assemblage (2011, 79–88). In conversely explaining how Nature writes, a position reviewed in our previous chapter, Kirby questions why "does Nature require a human scribe to represent itself?" (86). Such a view seemingly evokes Mead's non-hierarchical social to a greater extent than Latour.

A question that strikes me from Kirby's critique of Latour is that if, as is being argued now, social synchronization is a naturally immanent force, where does this leave the social representation of time, which the social constructionist time condition installs outside naturally real rhythms? Is there no representation of time, just time's reality? Rather than reducing our interpretation in this way, I propose that there *are* representations of time, and that such representations are inherent to time's reality, as per the following logic:

1. Time incarnates as the materializations of the real/world/natural.
2. Such materializations are intersubjectively/socially co-constructed.
3. Socially constructed phenomena represent the real/world/natural.
4. If, (1), (2), and (3), such materializations are of the real/world/natural *and* socially representational of the real/world/natural,
5. Then, time as materializations is simultaneously real and representational.

A further assertion from the above is that the reversibility of perceiver and perceived, which has been argued phenomenologically to be time, is also the reversibility of the representational and the real. Synchronization between incarnations of the real (including, but not exclusive to, human subjects) is not exclusively an arrangement that is wholly regulated by approximate constructions or representations of an inaccessible real/natural world-source. Subjects instead synchronize *as time*, enacting Being's self-synchronous, self-social, self-representation. The social construction of time is still defined here as a representational mode. The difference between my interpretation, and that inherent to the social constructionist time condition, is that in my model representation occurs from within, and as, the real of Being. This, as might be noted, is also the impression that has been developed of Being's self-theorization, whereby theory perceptually materializes Being as its representation and concurrent reality.

A critical characteristic of Merleau-Ponty's phenomenology that has emerged from its application to the context of theory's timing is the contestation that is engendered to the "reified," social constructionist opposition between naturally material, celestial time and socially abstracted, representational time. The real of social time is its naturality, whereby the social construction of time does not mark that which is contingently separated off from actual material nature, but rather where time's representation and reality manifest from the same objective source. With this universal, and therefore global, sense of temporality, the earlier appraisals of social time structures can be reconceived. Rather than a hierarchy emerging in which Filipino time or Caribbean time is seen to be aligned with a more natural state, what is argued here is that *all social time is natural time*.

A monistically natural mode of sociality, which in self-authoring is self-representing, coheres with my earlier accommodation of social constructionism within the real. Time, the internal differentiation of the real, concurrently produces and represents as temporalizing corporeal and theoretical bodies. A socially temporalized mode is an immanent perspective of time's entire nature rather than a localized approximation of a transcendent universal. The social construction or representation of time is Being's self-representation, dissolving representation and production into the one, simultaneous, worlding tempo. From this we can look beyond the interpretation that socially informed, cultural disciplines such as "philosophical theory" arrive late into an already originated, natural, real scene that they must then attempt to decipher. Instead, the dissolution of the reality of natural Being with the representation of natural Being into a singular process, marks theoretical endeavor as always already the real. Indeed, it could be rhetorically asked what, exactly, is the world ever theorizing or representing but itself, as itself, synchronously? That Merleau-Ponty's recalibration of the phenomenological method has framed and facilitated this appreciation of theory is, in my view, a most critical attribute.

As I have defined, synchronization refers to the perpetually co-constitutive becoming of inescapably synchronous subjects: Being's self-socialization. Because no subject preexists such intersubjective origins, synchronization is not restricted to an arrangement between already established subjectivities. Rather, synchrony includes a reference to the intersubjective simultaneity by which subjects become subjects. Mead's earlier appraised thesis interestingly discusses the relations between an organism and its environment according to similarly co-constitutive, co-distinguishable, intersubjective conditions of subjectivity, in which "the others and the self arise in the social act together" (Mead 2002, 178). This coheres with the timing that the phenomenological method brings to our attention regarding perceiver and perceived manifesting co-immanently and concurrently. To see is to be reversibly, materially seen.

Mead indeed complements inquiries into a concurrently local and global materiality of sociality via his discussion of an animal's physiological reaction when eating. This he describes not simply as an internal response to external objects. Rather, in recognizing that objects such as food constitute a subject's/animal's/organism's physiology and that this physiology simultaneously constitutes the world of objects, the animal emerges with/as the spatial/material/physical world of those objects (Mead 2002, 93). This singular, co-constitutive relation between physiology and the physical world (re-)produces the animal organism and the ecology/environment together. Or in Mead's terms, the event comprises "both the difference which arises in the environment because of its relation to the organism . . . and also the difference in the organism because of the change in the environment" (2002, 37). Mead's insight is that this co-emergence is sociality. That which emerges upsurges in more than one "structure" concurrently; "the emergent lies in both, and is what it is because it carries the characters of both at once" (98).

In connecting this to the phenomenological method that has been reconfigured by Merleau-Ponty, we can actually note that this ambiguity of corporeality, organism, and subjectivity evokes Heidegger's aforementioned contestation to the notion of finite bodily limits. For Heidegger, the organism does not occupy a separate "position in space" (Heidegger 1962, 420), but instead emerges as an aspect of "the equipmental totality" (1962, 420). It is from a singular, worldly whole that organism and environment simultaneously co-emerge as equipmental features of what Mead describes as "an ongoing living process that tends to maintain itself" (Mead 2002, 37). This process of becoming evokes the singularity and synchrony by which I have described Being's production of itself, whereby climate change materializations *world* the world, animating Being's changing perpetuity rather than indicating an end of Being. The human-as-body, a socializing, temporalizing, monistically plural, material space emerges with, and as, all other materialities/spatialities including those of climate change. This fleshed cacophony *is the social; it is time*.

This developing insight provides a novel phenomenological reconfiguration of how we might interpret the relation between humans and the advent of climate change. The earlier observed adversarial distinction of humans from the Earth's ecology is invoked if we reinvite to the discussion David Runnalls, who portrays human activity as *disturbing* "three quarters of the Earth" (Runnalls 2008, 20). The renowned theorist of environmental studies, William McKibben, provides an even stronger characterization on the same theme, arguing in *The End of Nature* (2006) that "true nature" has an existence independent from the human intrusion upon it. For McKibben, it is because of an essential conflict between humans and nature that the latter has become, as a result of human, cultural developments, an "artificial nature" (McKibben 2006, 115). Most forcefully, McKibben urges us to reflect on how nature became artificial not merely because humans arrived but because our arrival "deprived nature of its independence" (2006, 50).

In responding, we can return to Merleau-Ponty's phenomenological critique of Marx's "end of history." For Marx, as for McKibben, there is something about the present, social condition that is interpreted to have divorced humanity from its natural, material state, as well as corrupted nature's actual reality. Furthermore, for both Marx and McKibben, this separation between the natural and the social has been caused by human industrialization. Merleau-Ponty's reflection on Marx has been, consistent with his critique that the phenomenological reduction can never be "complete," that Marx neglects our ongoing, historically material, constitution. For Merleau-Ponty, there is no "end" of history.

I believe there is an application of Merleau-Ponty's contestation here that is also relevant to positions such as McKibben's, in which the notion pervades of a past in-itself, a natural/environmental state of pure primordiality. Rather than presuming that human materialization marks an externally imposed "end of nature," and with it the probable end of humans, we can define human-world relations more in terms of a self-responsibility. A responsibility in environmental care can in this way be portrayed not simply as something humans are ethically obliged to recognize and subsequently act upon, or as a responsibility to a natural Other that derives from the corrupting influence of the subsequent existence of humans. Instead, a responsibility to environmental care can be positioned as the perceiver's perpetual duty to a worldly, ecological, perceived self, an auto-responsibility that is inescapable. From this conceptual shift, just as the notion of social synchronization has been reconfigured, equally the interpretation of intervening *late* becomes problematic. To care for the environment is not to care for an inaccessible other that distantly precedes a social structure that has damaged it, but to care for the ongoing and always originating self-as-other.

While this chapter's engagements have not sought an alignment with any particular position regarding the recognition of, or the action toward, the reality of climate change, its argument does not preclude political concerns around environmental issues either. This work's politics are found in the interrogation of the separatist conditions identified within appraisals of human-ecological relations. In arguing for the common, rather than the secondarily oppositional, terrain of the real and the representational, an acknowledgment is demanded that the worldly "stuff/flesh" of climate change is the same worldly "stuff/flesh" that manifests political action to "deal" with such climatic change. The concern from climate discourses about the timeliness of our representations of, and interventions to, a changing worldly ecology would require a certain reconceptualization if this understanding of world-human, nature-culture, shared compositions was adopted. Furthermore, according to the implicated relation of theory *as* reality developed in the preceding debate, this phenomenological reflection that has been applied to the climate context could not be characterized as a philosophically futile or merely "academic" representation of physical reality. This might be the tendency when comparing my approach to the admirably intentioned, overt calls of environmentally geared scholarships to enact particular ways of arresting climate change. Conversely, in accordance with the key point that has been developed of the material actuality of any theoretical reflection, this chapter's thoughts, I argue, must be seen as similarly involved in materially real relations with, and as, worldly climates.

Despite the global synchronies proposed, I have noted that arriving late *is* an everyday, social reality. Humans do not always socially synchronize, people are left waiting for a missed bus, a meeting does have to be put on hold for tardy attendees, and there is a genuinely situated politics relationally differentiating how global populations manage lateness. These aspects of daily being do not seem to be consistent with the monistically synchronized world proposed here at all! It has earlier been revealed that in regions such as the Philippines and the Caribbean, protocols regarding lateness can appear to be relatively ignored. However, we should also note that if climate change is not arrested, it is argued that the parameter of lateness regarding collective responses to climate change could matter greatly to such populations, given how such regions would be some of the first to be affected by rising sea levels.[11]

Even though the social constructionist condition separating natural time from social time has been unsettled, we have reached a point that leaves the reality and actuality of lateness hanging. *Lateness still matters*. The all-encompassing, synchronous, materially social temporality discussed to this point has been described as monistic. A monistic ontology should be able to accommodate all manifestations of time, synchronous or not. While earlier, Adam's "out of sync" time-frames between humans and CFCs was recalibrated to recognize their mutually temporalizing synchrony, in the next chapter we will be considering how, even in maintaining all co-distinguishable relations of mutual synchrony, localized asynchronies are not to be dismissed. As it stands, we have attended to why the philosophies of time that have been engaged can explain monistic *synchronization*. The question that remains is how to retain the sense of a synchronous ontology in which natural and social times are not transcendently separated, while avoiding excluding the reality and possibility of the everyday experience of late or asynchronous relations.

Chapter Six

Methods of Accommodating Lateness

The Representation Inside the Real

> Time and space are fragments of the infinite for the use of finite creatures.
> —Henri-Frédéric Amiel, *Amiel's Journal* (2005 [1882], 80)

BERGSONIAN CRITIQUES OF SOCIAL UTILITY

The incarnation of social time has been recognized as nature's self-representational, self-distinguishable reality. Given the co-constitutive forms of this temporalizing, this self-distinguishing becoming will now often be described in this chapter as time's *co-distinguishability*. Through attending to what is social about natural time, or natural about social time, the distinction raised in early chapters between these realms by social constructionist tendencies has been interrogated. The myriad cultural forms of social time, and associated parameters regarding lateness, are not restricted to being contingent, social phenomena that simply represent from afar, a singular, natural, celestial phenomenon. Instead, via deconstructive and phenomenological reflections on the conditions of their respective perspectives on time, a monistic constitution of socially structured phenomena and naturally material phenomena has emerged.

Notions of lateness in our considered contexts have also been reconfigured. An engagement with body modification discourse has developed the insight that such modifications do not represent a culturally constructed or mediated violence that arrives late to an already established, natural corporeality. Rather, a modifying violence is the very condition of being bodied. Similarly, in terms of climate change, the interpretation that humans are responding too late to ecological change has shifted in order to appreciate a naturally materially synchronous temporality that is not separate from social or industrial rhythms. Such intervention into assumptions of secondariness has been extended to the distinction between material reality, and theoretical representation. A phenomenological critique of phenomenological considerations of perception and representation has positioned theory as Being's materializing self-reflection.

Nevertheless, in being navigated by this philosophical attention on a monistic, globalized synchrony, the reality of social lateness persists. People are still late to work, late to a meeting, and late for the bus. Lateness occurs, and it will continue to be a feature of your life! Recognizing the reality of lateness means that it needs to be accommodated within the model just developed of a global, material synchrony. If the temporality of co-distinguishability is *monistic*, then the question in this chapter will be whether our third philosophy of time, in reflecting upon presuppositions regarding time, is able to incorporate *all* occurrences of socialized temporality, even those that are seemingly not synchronous. If time is really monistic, then a critical appraisal must recognize not just time's originary or natural, global synchrony but also its localized and socialized asynchronies. This is a point that will remerge soon in Horkheimer's critique of the Bergsonian perspective that drives this chapter.

Given that synchronization is a binary opposite of the synchronization of lateness, if we follow the deconstructionist method encountered earlier, the identity and conditions of lateness should be present within synchronization. The reason for this, as Derrida has argued, is that one half of a binary pairing is as present in its "opposing, relational node," as it is in "itself." Accordingly, if we reflect on our discussion of social synchronization in the previous chapter, there we observed how everyday experiences of lateness are conditioned by experiences with the parameters of representational, clocked and calendared, time, which might also underpin two people synchronizing. Without some kind of division of time along a directional path, as separate minutes and hours or through an alternate, culturally sanctioned differentiation of points of time, experiences and adjudications of lateness are not possible. To be late is to fail to arrive somewhere or to do something by a certain point in time. Where time is divided into time points/states, lateness can manifest.

This sense that divided time states are the condition of social lateness does not have to be restricted to the mechanical, industrial representations of time on clocks and calendars. Our initial engagements in this book revealed the argument within forms of social science that not all societies recognize lateness to the same extent and that certain societies that do regulate lateness do not necessarily do so with the symbols of a twelve hour clock but rather via an adherence to the "ecological or material present." While the problems this has presented in marginalizing what are perceived to be more natural structures of social time have been explored, what I want to draw attention to now is that in any social environment, lateness is defined by a "when" that in turn is defined by spatial division. Whether this "when" takes the form of a separated minute marker on a clock, or of a certain position of the sun in the sky that is differentiated from its other positions, what manifests is that spatially divided states that are associated with temporal transition facilitate adjudications of lateness. Spatial division conditions the interpretation of "here but not there," or of "then but not now," either of which is often the hallmark of the socially late. Having acknowledged that the correlation between lateness and spatial division transcends any particular cultural context or representation, to reduce any potential verbosity in the discussion that follows I will often simply refer to the clock as evidencing this connection.

In order to attend to the reality of social lateness, we will therefore direct our attention to the notion of divided time states. This suits an examination of what awareness our third philosophy of time, Henri Bergson's, might be able to offer of the relation between:

1. lateness-conditioning divisible time states, and,
2. the co-distinguishable time states by which a monistic synchrony has been defined.

What this asks in particular of Bergson's philosophy is whether it can make apparent how divided time states are not outside the naturally materializing, monistic reality of time that has just been posited. This endeavor is bound up in a critical potential of Bergson's philosophy, re-encountering constructionist separations and asking whether "real time" is not simply a natural synchrony but also comprises socialized asynchronies.

The possible ramification for the kind of interpretation found in the citation from Henri Amiel that opened this chapter is a space and a time that are at once immanently infinite *and* transcendently finite. There is no more rigorous manner in which to explore this than by running the previous chapter's argument through one of the most prominent critiques of conceptions of material and "spatial time." It would indeed be remiss to posit that a reality of time can be materially spatial without engaging Bergson, whereby what will be interrogated is Bergson's demand that there are, conversely, "real" aspects of time that are *separate* from space. In our focus will be Bergson's reflection on social and philosophical representations of time according to spatial parameters and their supposed subsequence to, as well as distinction from, time's real/actual condition. As we move through Bergson's critique, we should be conscious of where his perspective might be situated in relation to our interrogation of the separatism that the social constructionist time condition requires between socially representational time and naturally real time. Philosophy students and scholars might find a particular appeal in the practicality that is illustrated in Bergson's work, as we apply it to questions of social lateness.

We will be expressly interested in the facet of Bergson's philosophy that manifests as perhaps *the intellectual* exemplar of the supposition of separated sources of times. In *Time and Free Will* (1960 [1889]), Bergson observes that humans conceive of experience spatially. The reason "we usually think in terms of space" (Bergson 1960, ix) is said to be identifiable in language. Human language descriptions impose the differentiating logic that seemingly reflects how things are distinct/divided in the materially spatial realm. For example, a rock is described as being in a different space from the dirt on which it sits. This tendency, however, can also conceptually divide otherwise "non-spatial" things, such as conscious states, installing "the same discontinuity as between material objects" (ix). Bergson later extends this point, changing his vernacular from the "imposition" of spatially conceptual language, to its "utilization," which suits "the requirements of social life" (128). Employing spatial representations facilitates social discourse, given the common, widespread assumption of ontological division between spatial things. Bergson reiterates this in *The Creative Mind: An Introduction to Metaphysics* (2007 [1934]), exclaiming "how much simpler" it is to use the spatial notions "stored up in language!" (Bergson 2007, 24). The point is that for Bergson, implementing spatial conceptions is socially pragmatic, whereby what "is spatial by nature has a social utility" (16).

Bergson's inquiry into why *all* experiential phenomena are "misconceived of" as spatially divided/discontinuous informs his concerns about our impression of time. Bergson demands that philosophy historically "misinterprets" time as only an extensive, spatial magnitude, whereby "when we evoke time, it is space which answers our call" (Bergson 2007, 4). The relation of extensive magnitudes to intensive states is central to Bergson's distinction of space from time and will be given our attention in the next section.

Before directing our focus there, however, I now flag the first of the reasons that will be detailed in this chapter regarding why Bergson's project can be conceived as one of "critical endeavour." Consistent with preceding characterizations of criticality in this book, this concerns his interrogation of structures that are so socially entrenched that they are rendered practically invisible. Bergson's intervention is going to occur in the context of the taken-for-granted "spatialization" of social and philosophical language, evoking an attention on the "reified social structures" that we have discussed in previous stages of this investigation.

My characterization of the criticality of this motivation of the Bergsonian reflection can be situated alongside recent engagements with his work. For Timofei Gerber, there is a "politically charged" direction in Bergson's "call to uncover the invisible structures and mechanisms that predefine the routes that we follow . . . a critical potential . . . even getting very close to Critical Theory" (Gerber 2017). Indeed, when specifically applying Bergson's critique of the spatialization of language to a form of literary analysis more closely aligned with the parameters of New Criticism, Byrony Randall uses Bergson's impetus to "contribute to a critical landscape where an emphasis on the spatialisation of temporality has resulted in a neglect of how temporal terms operate and circulate" (Randall 2007, 63). As we explore Bergson's place among already encountered philosophical destabilizations of tendencies deriving from what I identify as the social constructionist time condition, certain critical characteristics will again be positioned in accordance with the kinds of perspectives required by New Criticism.

INTENSIVE STATES ARE NOT JUXTAPOSED, EXTENSIVE STATES ARE JUXTAPOSED

Time and Free Will defines extensive magnitudes as measurable and comparable, whereas intensive magnitudes are not (Bergson 1960, 3). According to Bergson's commentary, space is extensive because it is divisible into distinct units (1960, 1–3). Collections of units, spaces, are also distinct from other collections/spaces and so can be measured and compared as greater than, less than, or equal to each other. Conversely, intensive states, "inner" magnitudes such as "joy or sorrow" (7), are not comprised of comparable, measurable, separate units. While a stronger or a weaker intensive state/sensation will seem analogous to the magnitude of space/extensity, it is Bergson's claim that there are not distinct, quantifiable units of something like happiness by which to measure it. This seems to be a fairly reasonable assertion from Bergson. How does one count or aggregate something like happiness after all?

Instead, conscious/intensive states of differing intensities are said to manifest via their co-constitution. Joy, Bergson notes as an example, (re-)produces the kind/quality of other conscious/intensive states, marking "qualitative alterations in the whole of our psychic states" (Bergson 1960, 10). Such co-implication means the distinction between intensive states is indiscernible. The phenomena of consciousness cannot be treated "as things which are set side by side" (1960, 8–9). Intensity is misrepresented, posits Bergson, when its co-implicated, qualitatively differentiated conscious states are divided into sections of the supposedly same phenomenon said to only differ quantitatively/extensively. In setting up "points of division in the interval which separates two successive forms of joy" (11), a change in the magnitude of the same intensive state is assumed, a quantitative alteration "of one and the same feeling" (11). However, according to Bergson's presentation, this transition is between different kinds/qualities of intensities and not a quantifiably comparable, extensive/spatial change.

It is worth pausing here to observe that Bergson's conception of co-constituting intensive/conscious states, which share a qualitatively transitional relation, evokes the characterization of time developed with Merleau-Ponty. There it was considered how time states co-distinguishably emerge from/as other time states. Intensive states for Bergson evidently share this co-constitutive/co-distinguishable ontology.

The characterization presented in our evaluation of the Merleau-Pontian co-constitution of time states involves material/materializing spatial/spatializing subjects/objects. Consequently, we will now clarify Bergson's position on what is spatial/extensive. Complementing his intensive|extensive, and consciousness|material, oppositions, in *Time and Free Will* Bergson recognizes "two kinds of multiplicity . . . one qualitative and the other quantitative" (Bergson 1960, 121).[1] Bergson explains a quantitative multiplicity with reference to counting. We are told there must be the assumption of something homogeneous about each of its individual units so that they can be counted or quantified together. Nevertheless, he reminds us that the units must be distinct, "otherwise they would merge into a single unit" (1960, 77). Bergson illustrates his point by discussing the counting of sheep. Each sheep is unique he notes, spatially distinct in a paddock, yet in counting them we "neglect their individual differences" in order to "take into account only what they have in common" (76).

What countable units also have in common for Bergson is simultaneous presence, which counting requires for number/magnitude to accumulate/increase. Given that these homogeneities must be distinct, each is spatially set "alongside each of the new units" (Bergson 1960, 77). It was earlier observed that, conversely for Bergson, intensive states are not set aside each other but are co-implicated states. Furthermore, it was acknowledged that this co-implication evoked the co-distinguishability of states of time developed in our last chapter. As we will now see, that was no mere passing evocation.

According to Bergson, the human error concerning real time (duration/*durée*) occurs when homogenous representations of comparable, measurable points are deemed to be time's *only* constitution. Think here of a clock or a calendar. By lining up and counting successive moments of duration, Bergson notes that time seems to be only "a measurable magnitude, just like space" (Bergson 1960, 104). In *The Creative Mind*, Bergson attributes this to our intellectual desire to know locations, where "fixity is what our intelligence seeks" (Bergson 2007, 5). We can recall here Bertrand Russell's observation from chapter 2 concerning the human desire for permanence. In counting the spatial traces of fixed points, for Bergson "all the intelligence retains is a series of positions: first one point reached, then another" (2007, 5). The homogenization of discontinuous points facilitates their quantification. That is, we conceptually divide, and then count, time. Such a quantifiable multiplicity of discontinuities is, of course, Bergson's conception of space/extension presented in *Time and Free Will*. From this, Bergson concludes that time, understood only as an extensive "medium in which we make distinctions and count, is nothing but space" (1960, 91). We have seen that contrarily for Bergson there is no separation between intensive states.

As outlined in this chapter's opening, this definition of spatialized time as a *divided* medium marks the relevance of investigating its accommodation in a monistic ontology of co-distinguishability. This informs our study of whether the conditions of the everyday, social reality of lateness are not excluded from our sense of time's reality. Spatialized time, divided time, of homogenous, comparable clocked points, or of differentiated positions for the sun, conditions the regulation of social synchronization and lateness. Here I have argued that it is according to spatialized representations of time, what Bergson describes as a series of mutually juxtaposed points, that people can be judged as late.

Bergson conceives of time's conversely intensive state as time-as-duration/*durée* and differentiates it in *The Creative Mind* from time's spatial representation via three analogies. The most straightforward of these describes a piece of elastic that when stretched becomes longer. This evokes time-as-duration for Bergson if we focus on the action of stretching instead of on the resultant stretched elastic (Bergson 2007, 138). If we simply focus on the result (as indeed we do when we quantitatively represent time), our impression of time is, rather strangely, of a static space of divisible points. Conversely, the stretching for Bergson indicates that no two points of intensive time can be aggregated because each point, each time-state, is produced by qualitatively transitional relations with new points or time-states that emerge (2007, 137). States of time in this durational regard thus differ from spatial states because "duration excludes all idea of juxtaposition and reciprocal exteriority" (138).

Bergson's willingness to exclude ontological divisibility from co-permeating time-as-duration indicates the restriction he installs at the very point where our work in this chapter, in exploring the accommodation of divided space-as-time states with co-distinguishable time states, commences. We should recall here that for Bergson, the social, spatial construction of time, is seemingly separate from the worldly phenomenon that it represents. Conversely, the argument developed in our previous chapter is that a monistically social construction is inherent to worldly phenomena's reality. Nevertheless, as part of this undertaking we will later ask whether Bergson inadvertently relies upon a similar singularity, rather than an opposition of reality and representation, in order to secure his argument.

I posit for now that Bergson exhibits certain tendencies that are consistent with the social constructionist time condition via his argument that time, when represented extensively/spatially, is a series of divided/discontinuous points. This is because, for Bergson, any such socially impelled representation is an imperfect, partial version, "taken from a certain point of view" (Bergson 2007, 135). These notions of imperfection mirror the inaccessibility with which we have seen constructionist theory describe the relation between the real and representation, and is the basis upon which Bergson criticizes Kant's refutation of the possibility of absolute knowledge.[2] Bergson's position is here congruent with the social constructionist time condition in terms of its polarization of the truth of reality from what are said to be the imperfect contingencies of social representations of reality. While this is not a common focus of readings of Bergson, it is in concert with recent observations of the distinction he posits between nature and culture.[3]

Converse to the socially spatialized reflection on, and division of, time, Bergson describes the lived experience of time-as-duration as indivisible and "intuitive" (Bergson 2007, 138–39). This intuitive character of time for Bergson is both synonymous with duration *and* a method for accessing duration. When synonymous with duration, intuition simply lives/experiences the flow of co-constitutive, durational states. As a method, however, intuition opens beyond this experiential state, toward an awareness of what Deleuze describes in *Bergsonism* as "the conditions of experience" (Deleuze 1988, 27). Intuition in this mode considers the heterogeneous relations of co-constitutive, qualitatively relational states, which as we have seen is Bergson's definition of states of real-time-as-duration. Bergson consequently notes in *The Creative Mind* that "to think intuitively is to think in duration" (Bergson 2007, 22). The complementary assertion is made in *Time and Free Will* that intuitive consciousness is conditioned by the co-implication, rather than the socialized conceptual separation, of states of duration (Bergson 1960, 100).

Deleuze's *Bergsonism* articulates that by considering how states transition qualitatively, "intuition has become method" (Deleuze 1988, 32). Intuition as method is *not simply duration.* Nevertheless, intuition as method manifests through the conditions of duration, opening consciousness to duration's qualitatively transitional differences in kind (rather than merely differences in degree/quantity). Such states are consequently, as Bergson describes, "all very different from one another" (Bergson 2007, 156). In considering intuition as method, I argue that we develop here previous chapters' insights regarding what is implicit to, rather than representationally estranged from, reality about method and theory. Moreover, given that for Bergson intuition as a method is not restricted to a subsequent reflection upon an already originated, temporal reality, this seemingly problematizes the characterization of a human production or activity arriving late on the scene of a naturally originating process. Rather, intuition as method demands that the human functioning of time coincides with, and as, what Bergson positions as "real time." In this regard for Deleuze, if methodological intuition did not coincide with this durational process, it "would not be capable of . . . the determination of . . . genuine differences in kind" (Deleuze 1988, 33).

Summing up preliminarily, time for Bergson is experienced *both* as quantitatively different, spatial degrees *and* as qualitatively different co-constitutions. One's experience of time is comprised by the everyday social requirements of spatial, measurable, divided discontinuities, *as well as* by the internality of non-spatial, durational co-distinguishabilities. Only the latter mode is real time. While the argument of our last chapter, which does not separate social time from real time, thus disagrees with Bergson on the status of social time, Bergsonian time-as-duration shares with that argument the impression that time involves co-constitutive upsurges that emerge distinguishably from/as each other.

In addressing intuition, we have seen Bergson describe these co-constitutive states of time-as-duration as *successive*, in that "states of consciousness, even when successive, permeate one another" (Bergson 1960, 98). This raises an interesting point; how can a philosophy describe states of time as being perpetually involved in/as each other co-constitutively, yet also describe such states as successive? Do we not describe succession as when a state replaces rather than simultaneously coexists or co-permeates with preceding states? If time-states co-permeate (coexist) *and also* succeed (replace) each other, is there a concurrent *simultaneity and succession* in the time-states of Bergson's argument? Why would this even matter? Well, as we will now reiterate, this is how I have phenomenologically characterized materially spatial time(s). Furthermore, these dual parameters lurk in the conditions of lateness.

SIMULTANEITY AND SUCCESSION: BERGSON'S SPACE

Rather than past and present simply succeeding one after the other, duration for Bergson "forms both the past and the present states into an organic whole" (Bergson 1960, 100).[4] This correlates with the succession already noted of the perpetual (re-)emergence of past(s) and/as present(s) and/as future(s). The continuity of this succession is in contradistinction to the discontinuity of space(s) for Bergson, given that, as stated in *The Creative Mind*, consciousness of/as time-as-duration "grasps a succession which is not juxtaposition" but shifts from within as "the uninterrupted prolongation of the past into a present which is already blending into the future" (Bergson 2007, 20). Contrarily, Bergson demands in *Time and Free Will* that where "we should find only space" there is "nothing but simultaneities" (Bergson 1960, 116). Spatial simultaneities are juxtaposed, set out in line (1960, 102, 115, 226), without "duration nor even succession" (120).

Despite the "blending" of successive states, succession for Bergson *cannot* incorporate simultaneity. The heterogeneous relations of time-as-duration's co-permeating states demand that "not one of its parts is still there when another part comes along" (Bergson 2007, 2). Because duration is always in qualitative re-production, it "eludes mathematical treatment" (2007, 2). Such mathematical or scientific reflection returns us to spatial time, which freezes what has already occurred into symbolic, static, *objectively simultaneous* representations that are comparable and combinable. These homogenous representations also anticipate how time *will* manifest. It can be recalled that in order for homogeneous, simultaneous entities to avoid dissolving into each other indistinguishably, they must be mutually excluded via spatial intervals. This mutual exclusion differentiates the simultaneity of separated/divided/juxtaposed spatial states from the co-constitution by which states of time-as-duration perpetually succeed.

It is from this differentiation of simultaneity and succession that we can further situate the critical elements of Bergsonian perspectives regarding time. This will lead us to a brief analysis of Bergson's conception of art, which in how it informs aspects of literary and poetry studies will contextualize the aforementioned relationship between Bergsonian perspectives on time and New Criticism. Because time-as-duration for Bergson is an *intuitive* mode, and is not comprised of simultaneously homogenous states, there is no possibility of "going back in time." Intuitive states are, as we have seen, always already becoming qualitatively different states, whereby as Bergson notes in *Creative Evolution* (1911 [1907]), "consciousness cannot go through the same state twice" (Bergson 1911, 5). Alternatively for Bergson, the spatialization of time is comprised by homogenous states in a line, whereby spatial time has a predictability based on the repeatability of any of its states. This repeatability of mathematically or scientifically measurable, spatial time conditions the capacity to predict the future. Such prediction is based on applying a perception of the pattern of the past to an anticipation of what will occur, or in Bergson's words, where "to foresee consists of projecting into the future what has been perceived in the past" (1911, 6).

This feature of spatial phenomena is exemplified for Bergson in geometry, where a truth to its general order is said to preexist any of its individuations. That geometric laws can mathematically predict the constitution of objects that have not yet been produced identifies this (Bergson 1911, 7). Conversely, art is said to represent the *unpredictability* of the time-consciousness of duration and intuition. With a completed production of something like portrait art, Bergson argues that "even with the knowledge of what explains it, no one, not even the artist, could have foreseen exactly what the portrait would be, for to predict it would have been to produce it before it was produced" (1911, 6). The point for Bergson is not that art is without order, given that signification is still present in the art that is produced.

Instead, art manifests as an order of unpredictability, by lacking an order that preexists its realization. As Mary Gillies notes, this broadens the scope of art for Bergson from its mere products, to be "more than a finished poem, painting, or symphonic score; for Bergson, art is the experience of these things" (Gillies 1996, 20).[5]

From this perspective we see that critical contexts open to invite Bergsonian theory. When Ewa Thompson says of the field of New Criticism that it was the English critic and poet Thomas Hulme who inadvertently became "one of the fathers of a critical trend" (Thompson 1971, 39), part of the reason can be said to be attributable to the inspiration that he derived from the intuitive order of art found in Bergsonian philosophy. Thompson focuses on various aspects of Hulme's works, including his critique of Romantic poetry, to develop this characterization of his integral "part of the New Critical traditions" (1971, 40). Complementing this, she notes the influence of Bergson in shaping Hulme's perspectives on poetry and aesthetics (66–67). In terms of how this manifests in Hulme's commentary in *Speculations* (1936) on Bergson, I am inclined to follow Donald Davie's 1955 assertion that "it is probably due to Hulme that much modern criticism is Bergsonian" (Davie 1976, 6). Davie, in referring to the pivotal work of John Crowe Ransom in the inception of New Criticism, notes that when "Ransom writes that the poet's is a world of 'stubborn and contingent objects,' with a sign up, 'this road does not go through to action; fictitious,' he is writing quite in the Bergsonian spirit" (1976, 6–7).[6]

While Hulme is adamant that Bergson has "not created a new theory of art" (Hulme 1936, 143), Bergson's conception of intuition significantly informs how Hulme's critique of poetry develops. In identifying two reasons for this, Hulme first notes Bergson's demand that art, instead of being a part of the spatially appraisable and actionable world, is a "more direct communication of reality" (1936, 146). Reality for Bergson here refers to the unpredictable order by which the painting manifests, or in Hulme's words, the "flux of interpenetrated elements unseizable by the intellect" (146). In evoking Ransom's, Bergsonian-inspired, New Critical perspective regarding the difference between modes of intuition and the socialized necessity of action, Hulme is bothered that "man's primary need is not knowledge but action" (147). It is according to Hulme that Bergson attends to this distinction when conceiving of the intellect. Intellectual presentations of things, said to be oriented scientifically, spatially, and perhaps most importantly, socially and pragmatically, do not reflect the way that "we may most thoroughly understand them, but that we may successfully act on them" (147). The separation of reality (intuition) from social constructions of reality required for the logistics of action is evident.

Art in Bergson's philosophy, according to the endorsement of Hulme, is not the construction of actionable, or intellectually intelligible, things but operates as a dismantling of the normalizing tendency that blocks our intuitive mode (Hulme 1936, 174). It is by applying this perspective to an appreciation of poetry that Hulme, in "Romanticism and Classicism," duly describes poetry as "a compromise for a language of intuition that would hand over sensations bodily" (Hulme 1994, 70). My reading here is the very Bergsonian point that intuition does not have a predefined structure or language, for if it did it would take on the homogenous qualities of a mode or geometrical template designed for action. Hulme says as much in *Speculations*, arguing that language excludes what "really exists and substitutes for it a kind of stock or type emotion," the result of this being that "most of us never see things as they are but see only the stock types which are embodied in language" (Hulme 1936, 166). For Hulme this characteristic of language is indicative in some forms of prose, as a dead end "train that delivers you at a destination" (Hulme 1994, 70). The finiteness of prose and language are here distinguished from the verse of poetry, in which for Hulme one becomes *involved*, stating that "all poetry is an affair of the body—that is, to be real it must affect body" (1994, 242). It is from this approach that Hulme is said, as evidenced in Thompson's earlier account, to have inadvertently developed a method that would inform New Criticism's approach to literary criticism. Via Hulme's declaration that "images in verse are not mere decoration, but the very essence of an intuitive language" (70), the connection with Bergson's conception of experiencing art is readily apparent, as is the inspiration this would offer New Criticism in terms of its focus on direct textual analysis. A later noteworthy commentary on this point is provided by Paul Edwards, who attributes the appeal of Hulme for New Criticism to Hulme's comparison of the critical value of Bergson's theory of art with his own theory of poetry. Edwards draws the connection that "what is at the root of both is a desire to make vivid, to avoid the merely habitual and the settled grooves of customary perception" (Edwards 2006, 31).

While this commentary explores the coherences between Bergson's philosophy and the parameters by which a theory comes to be defined as critically navigated, a competing voice to such coherence can be found via Horkheimer's Frankfurt School perspective. For Horkheimer, there is an aspect of Bergson's dual definition of time consciousness that is typically metaphysical rather than critically atypical. In one regard, Horkheimer endorses the Bergsonian characterization of time-as-duration as perpetual, qualitative change, given that this correlates with his own conception of history, stating that Bergson's "foundational theme, real time, is a central category of any thinking of history" (Horkheimer 2005, 10).[7] In another regard, however, Horkheimer criticizes the intuitive mode of Bergsonian duration, in that according to Horkheimer it presents succession-as-change as time's infinite essence, of "infinite transformation, of eternity" (2005, 16). Horkheimer conversely wants to recognize that humans have finite aspects, that they are situated in history in specific ways to the exclusion of other ways, and where "talk of existence as temporal means instead that this doesn't endure, but is finite . . . it ages and passes away—not merely that it changes" (16). This evokes my earlier detailed concern that if Bergson's philosophy of time is, in my estimation, to be characterized as a critical theory of time, it will need to accommodate not just a real time of originarily co-permeating, synchronously changing states but also localized and socialized asynchronies.

It is through the theme of lateness that I come to both disagree and agree with Horkheimer's critique of Bergson. In disagreeing, I note that via Bergson's incorporation of the socially normalized, spatial motifs of time, an experience of this "finite" version of time *is* positioned by Bergson as very real for us. Bergson's actual concern is not that this finite mode of mutually juxtaposed time states exists, but rather that we often confuse it as being all that time is. Indeed, in noting the reality of this kind of time experience, we can observe my consequent point that the mechanics of lateness are conditioned by this finite or spatial mode of divided points.

Nevertheless, I also agree with Horkheimer that Bergson does seem to present "qualitatively intuitional" time as infinitely and eternally "real or essential time," contrary to its social spatialization. This perhaps explains in part why I have the impression that while Bergson's methodology appreciates both divisibly separate and distinguishably co-permeating states of time, it does however rely upon the social constructionist's model observed in earlier chapters. This is where there is an internal reality to what is true about time and an external fabrication about how it is socially structured and spatially lived.

Where we left the previous chapter, a *simultaneously* co-productive *and successively* co-preserving materializing/spatializing/corporealizing-as-temporalizing emerged. The world's materializing was seen to be conditioned by its reversible, ambiguous, concurrent subjects and objects presented as Being's simultaneous representations and realities. These temporalizing upsurges have been posited as also a novel/new/originating succession consistent with the logic engaged in Mead's conception of time, where time states reversibly, perceptually co-originate.

We see that Bergson also correlates material space with simultaneity. However, Bergsonian simultaneous space/extensity apparently lacks the co-constitutive, co-permeating attributes by which we reconceptualized synchronization. The body-as-materializing-space is a site of interest here. *Time and Free Will* declares that time is misconceived when it is represented as moments "external to one another, like bodies in space" (Bergson 1960, 107). The key point of difference between our theses is that pure "externality" is not a characteristic of the body that has developed in our discussions of previous chapters. Observing this calibrates this chapter's direction. Bergson's *simultaneously divided/juxtaposed* spaces, what I identify as the condition of social lateness, are not yet accommodated within what has been positioned as the inescapable, primordial reality of time's *simultaneously co-constitutive/co-distinguishable* conditions.

Consequently, we will temporarily set aside discussions about space/materiality to focus on *simultaneity*. Can the sense of Being's monistic temporalizing that we have developed, in which all time states, even past and present, manifest co-constitutively and simultaneously, accommodate simultaneously divided time states? If this co-constitutively synchronous temporality is, in fact, *monistic*, the representational mode of simultaneously divided past and present time points needs to be included and explained rather than excluded. Excluding forms of time is a characteristic of the social constructionist time condition that we have destabilized, whereas I recognize that even simultaneously divided time-states have an experiential reality.

Here we jump to Bergson's later conception of time in *Matter and Memory*, which concerns time's relation to memory, a realm we have already encountered via Augustine. Bergson differentiates three processes: pure memory, memory image and perception. Perception, we are informed, is not simply an instantaneous, present representation of an object. Rather, perception is "impregnated with memory-images" (Bergson 2004 [1908], 170). Memory images come from what is described as "pure memory," and importantly for the current stage of our inquiry, it is the relation between perception and pure memory that indicates how present and past relate. By incorporating considerations of the pragmatics of everyday, social experiences of lateness, it seems appropriate to be exploring the topic of memory. A certain temporal synthesis is required in order to avoid being late, in which one anticipates in the future the memory of certain contracts made in the past.

The Bergsonian method for outlining a theory of memory involves the condition that when remembering, we "detach" ourselves from the present in order to reexperience first the past generally, and then specific regions of the past. Bergson likens this to the "focusing of a camera" (Bergson 2004, 171). The more that memory comes into view determines how "its outlines become more distinct" (2004, 171). Importantly, while this process "tends to imitate perception," memory is not a weakened perception, nor "an assembly of nascent sensations" (179). This is the interpretative error that Bergson attributes to "associationism." In separating memory/past from perception/present, associationism substitutes for the heterogeneous flow of time-as-duration a "discontinuous multiplicity of elements, inert and juxtaposed" (171). Conversely for Bergson, while memory and perception are radically different processes, they are *not* actually polarized. Indeed, memory-images participate in perception as the "dim nucleus" of vision (172).

Bergson's key demand is that interpreting memory as weakened perception commits the same error as spatializing intensive/conscious/durational states. This is because it defines past (pure memory) and present (sensation/

perception) as different only by degree or "magnitude" (Bergson 2004, 173). We have seen for Bergson that conversely, states of time-as-duration differ qualitatively, and it is by this character that he describes the relation of past/memory to present/perception. Nevertheless, they do operate in tandem, a matter that *Matter and Memory* discusses via his renowned illustration of the inverted cone. The base of the cone, which "remains motionless" (2004, 196), represents the "pure memory" of the past. Underneath this is the point of the cone, as the consciousness of present action. Bergson argues that as the perceptual present appeals to memory, memories participate in what the present perceives (197). Memory emerges as a utility of the perceptual present, noting the "utilitarian origin of our perception" (206).

Because the present calls to the past, and not the other way around, Bergson calls the present the "actually lived," or the "active" consciousness (Bergson 2004, 181). Conversely, the past-as-pure-memory is defined by its "latency" and "radical powerlessness" (2004, 181). Bergson indeed describes the memorial past as "detached from life" (179). This is an intriguing development as the notion of a "powerless, lifeless detachment," for any aspect of time, seemingly contradicts Bergson's point that the perceptual present and the memorial past are *not* entirely divided or separated.

What I wish to bring to our attention here is that Bergson's notion that if states are simultaneous then they must be divided/"detached" is not necessarily the issue. Indeed, the existence of simultaneously divided states is what we are ultimately exploring regarding their potential inclusion in, rather than exclusion from, what has been positioned as the monistic co-distinguishability of real time. What is an issue, however, is that for the divided/"detached" past to also be considered to be within this "real time," it cannot be characterized as being any less involved in temporalizing than the present. Such a distinction between past and present is, however, something Bergson posits in describing the past as latent and detached. Given the converse singularity by which Merleau-Ponty characterizes temporalizing structures, a reincorporation of his critique of not merely the phenomenological method, but now of Bergson also, could assist here.

Merleau-Ponty is critical of Bergson's conception of the present-past relation when exploring analogous questions concerning language. In *Phenomenology of Perception*, Merleau-Ponty argues against Bergson's interpretation that a speaker requires a separate thought-representation of a word before being able to say it. Instead, for Merleau-Ponty, the thought of the word *is* the speaker's speaking the word; "his speech is his thought" (Merleau-Ponty 1962, 209). This is consistent with Merleau-Ponty's sense of an experiential, phenomenological consciousness, where it was observed that a subject's knowledge of time is not simply a consciousness of, or a familiarity with, a predefined reality or representation but is attributable to every subject's real-and-representational, reversible, self-consciousness *as* time.

In articulating how consciousness is directly existential/experiential, and not simply indirectly representational, Merleau-Ponty contests the Bergsonian argument in which the conscious present draws upon a radically distant, latent, pure memory. Again this coheres with our previous inquiries in which it was noted that Merleau-Ponty's phenomenology, contrary to the Derridian accusation, does not hierarchize a transcendental present that can return to a fixed origin. Merleau-Ponty states that there is not a memorial storehouse in which consciousness delves to find a relevant word and "retain some 'pure recollection' of the word, some faded perception" (Merleau-Ponty 1962, 209). His critique here that Bergson defines memory as "faded perception" is not as straightforward as it seems though. This, after all, is the very interpretation against which Bergson argues. We have seen that for Bergson, what manifests when perception "calls" memory differs radically from what memory was, rather than becoming a stronger (or less "faded") version of the same phenomenon. Nevertheless, let us consider how Merleau-Ponty's interpretation informs what seem to be valid concerns about Bergson's possible separation of the perceptual present from lifeless, powerless pasts-as-memories.

The past-present division of what Merleau-Ponty describes as "Bergsonian dualism" (Merleau-Ponty 1962, 209) is interrogated by arguing that recollection is more originary than a purported search of far-flung, powerless memories-as-pasts, providing them with greater definition. Merleau-Ponty insists that "to remember is not to bring into the focus of consciousness a self-subsistent picture of the past" (1962, 26). This contestation to a "self-subsistent past" is consistent with how in this book the notion of a past in-itself has been problematized. While we have seen Bergson argue that the perceptual present summons from pure memory what is of "utilitarian interest," for Merleau-Ponty, one's entire past is always already implicated in/as perception, as a "synthesis of apprehension [which] links me to my *whole* past" (486; my emphasis).[8] Merleau-Ponty's point here is consistent with the argument we have developed that *all* time-states are co-constitutive.

This discussion in *Phenomenology of Perception* examines Bergson's point that memory-images are retrieved by the present to serve a particular sensory-motor utility/interest. The ramification of this retrieval process with which Merleau-Ponty is uncomfortable is that the present is only an impetus for these memories to be "relived." The memorial origin remains transcendently past as pure memory. Merleau-Ponty's argument that the Bergsonian past-present relation is one of mutual exteriority is most overt in *The Incarnate Subject* (2001 [1968]). This text observes that Bergson installs two distinct realms—one perceptive, the other memorial—which must then speak to each other. It is via the body that this communicative "junction" manifests (Merleau-Ponty 2001, 91). Remember, the body for Merleau-Ponty is not something that simply perceives. Instead, the body conditions Being's self-perceiving, self-constituting upsurgence. Merleau-Ponty is adamant that conversely "for Bergson, perception is not constituting" (2001, 89). The Bergsonian body is said instead to be divided between a sensori-motor perceptual presence *or* an image in pure memory (91). While Bergson argues that in lived experience these realms are inseparable, Merleau-Ponty is not convinced; "Bergson fails to establish the articulation between the two levels: pure percept and pure recollection" (91). Recent appreciations of Merleau-Ponty's differentiation from Bergson on this point are encapsulated by Ted Toadvine, who affirms that "Merleau-Ponty's critique of Bergson focuses" on this issue of supposedly prior purity (Toadvine 2007, 22–24).

If divided states of time are not conceived as real, primary, or powerful, the capacity of the philosophies of time being engaged in this book to destabilize separations of pure or real time, from subsequent representations or consciousnesses of such time, is seemingly hampered. For Bergson, if the body's sensori-motor perceptuality is separate as the present, then aspects of time-consciousness elude it. Merleau-Ponty concurs that when Bergson discusses "the body, he will leave consciousness of time out of his consideration" (Merleau-Ponty 2001, 96). Indeed, in *Matter and Memory*, Bergson describes the perceptually present body as utilizing memories imaged from the past, without being implicated in that realm, whereby "pure memory . . . interests no part of my body" (Bergson 2004, 179). Merleau-Ponty condemns this separation of memorial past from bodily present as lacking the passage required for time. Such a body can only be, as Merleau-Ponty states in *The Incarnate Subject*, "a means of actualizing the past," and restricts "the body as a present existent rather than a temporal reality" (Merleau-Ponty 2001, 96).

The body, as conceived via Merleau-Ponty, marks the ambiguous divergence of the flesh of the world from itself, constituting consciousness of/as flesh in general. Fleshy existence is fleshy consciousness, where such "existence . . . always implies conscious apprehension" (Merleau-Ponty 2001, 104). Conversely, Bergson, according to Merleau-Ponty, installs a subsequent "call" from substantial present, to ideal past, a "late" call it could even be said, where the consequent memory image that relays between the two develops time-consciousness. This is not the primordial consciousness of Merleau-Ponty's corporeality-as-flesh, nor of the sense through which we have guided ourselves of the all-proliferating, all-writing, all-theorizing, all-temporalizing body. Merleau-Ponty duly suggests to Bergson that "instead of *placing* in the world *seeds* of consciousness and instead of *leaving* in consciousness *traces* of materiality, he should have grasped consciousness as history and proliferation" (2001, 106; author's original emphasis). As noted in last chapter's discussion regarding the ambiguity of flesh, Merleau-Ponty does not reduce all aspects of fleshed perception to the materiality of the present. Rather, the generalized flesh conditions what is presently common to the materializing of materiality.

There is a symmetry between this style of Merleau-Ponty's critique of Bergson by Merleau-Ponty, and my argument in chapter 4 that the structuring-structured, constituting-constituted, co-production between subject and/as social of Bourdieu's sociology does not attribute the subject's productive involvement to the fundamental *matter* of their existence. Bourdieu requires the subject to undertake "*ways* of being," to action/do in order to structure and be structured socially. My concerns regarding Bourdieu deploy a similar logic to that which underpins Merleau-Ponty's accusation that simply being/existing is not a condition of knowledge for Bergson. For Merleau-Ponty, as early as *Phenomenology of Perception*, existence is primordially knowledge of/as Being, requiring no search for consciousness with "my body, nor with time, nor with the world, as I experience them in antipredicative knowledge, in the inner communion that I have with them" (Merleau-Ponty 1962, 82). Merleau-Ponty demands that Bergson, conversely, postulates an intellectual knowledge in which there is division/juxtaposition—a gap, a break, a loss—to be overcome. As with the social constructionist time condition that has been engaged in previous chapters, which characterizes representation as directed toward an inaccessible, real world/Being, Bergsonian intellectual knowledge according to Merleau-Ponty's critique requires one to "find a passage between a being who knows nothing and a knowledge cut off from this being" (Merleau-Ponty 2001, 110).

The heuristic nature of our incorporation of Merleau-Ponty's critical reflections of Bergsonian philosophy should be noted. The dialogue being developed is not designed to take the "side" of either. Rather, we are opening this discussion in order to consider how critiques between philosophies of time can be managed in order to develop an original inquisition of the social constructionist time condition's separatisms. Merleau-Ponty's reading of Bergson affirms the conception that if states are divided/juxtaposed, then one such state (the past), is closed off from the "real" of time's ontology. The conditions of lateness here are still a constructed, external representation of time's reality of co-permeating/co-distinguishable states. Reconsidering the timing between divided states from Bergson's perspective is accordingly appropriate, given his different interpretation (from Merleau-Ponty's reading) of how his work relates the past to the present. In undertaking this kind of reversible examination, we also become more acquainted with how critical interplay (re-)originates texts, where theses from different eras manifest co-constitutively, just as such theories might themselves posit of past and present states.

BERGSONIAN PRESENT-CENTRISM RECONSIDERED: MERLEAU-PONTY'S CRITICAL TRACES

It is Bergson's actualization of the past *in the present* that as reviewed bothers Merleau-Ponty. The reading here is that Bergson characterizes the past-as-memory as separate/detached/static traces from which the active present draws, producing "time out of a preserved present, evolution out of what is evolved" (Merleau-Ponty 1962, 482). Rather than an all-powerful, dynamic present, retrieving a latent, passive past trace, Merleau-Ponty's interpretation is of an always originating trace, articulated in the "Temporality" chapter of *Phenomenology of Perception*. The trace for Merleau-Ponty is not a finalized, purely memorial past from the viewpoint of the present but rather is a significance of the past that is always rearticulating (1962, 480).

Merleau-Ponty asks us to consider a carving in a wooden table (Merleau-Ponty 1962, 479–80). Recognition of the carving is only possible because one has a past sense of carving and tables. The past participates in conditioning present perception. However, the past is not something fixed in the reservoir that is Bergson's "pure memory," upon which one transcendently reflects. Instead, the perpetually originary nature of the past-as-"trace" is conditioned by the unreflective mode of experiential being. One's trace of what constitutes a carving and tables is always becoming something different. This logic was encountered in Merleau-Ponty's critique of the phenomenological method, where by always perceptually being-in-the-world, experiential phenomena can never be completely representationally "reduced" via reflection. Any reflective moment is conditioned by/as our "unreflective" being-in-the-world. We cannot parenthesize being-in-the-world, nor put it out of play completely, in order to reflect upon it. Or simply, the present from which one can never be extricated is always already co-constitutive with/as an emergent reflection/past.

We have addressed concerns about describing a state as *un*reflective, noting Merleau-Ponty's clarification that simply by being one manifests *as* consciousness. If reflection does not reduce Being to a series of knowable, in-themselves moments but rather participates in and constitutes that upon which is being reflected, then there is not a preexisting past that has previously been present to be recalled in its eternal, identical fixity. The reflective moment co-constitutes new, originary pasts, a primordial, phenomenological reduction that "constitutes for it [reflection] a kind of original past, a past which has never been present" (Merleau-Ponty 1962, 282). As seen via our engagement with Mead, this is the past that has "never been present" because it becomes the past that it is, with, and as, this present(ing). The lived, "unreflective" present is not simply the all-powerful condition for the originary past, as Merleau-Ponty identifies in Bergson's philosophy. Rather, the originary past, a trace/sense of the past-as-present, is always actively differentiating via/as the present perception with which it manifests. From this, any sense of secondariness, particularly within constructionist perspectives that deflect the representational mode away from its natural ground, is problematized. While in this chapter we are actually looking to incorporate the temporally divided states that condition social lateness, the hope is that Bergson's philosophy can do so in a manner that does not require compromising them as powerless. Achieving this would, I believe, represent a critical aspect of Bergsonian theory by not just recognizing a real time that is globally synchronous but also by intervening in a manner that is able to liberate localized and socialized structures of asynchrony from characterizations of being less-than-real time.

By considering the temporalizing power or productivity of the past/trace, our conception of primitiveness shifts. Typically, the primitive is comprehended as a present that eternally preexists, consistent with appraisals of an originary, natural time-source separate from the contingent, social constructions that follow. However, as developed during our previous studies in this book, what is primitive is a past (nature) that has never actually been present. The past was never simply, previously, present in-itself, but is always co-constitutively emerging. This is the primacy/primitiveness of the subject's "un"reflective gesture with/as the world. No time state, no one of our time states, neither present nor past, is fabricated by the others from without according to Merleau-Ponty, in that "no one of time's dimensions can be deduced from the rest" (Merleau-Ponty 1962, 492). Rather, what must be recognized is that our temporalizing marks our "primitive alliance with the world" (1962, 493). This is the basis of Merleau-Ponty's contestation to the hierarchization of the present over the past that he interprets in Bergson's theory and speaks to our earlier interrogations of the notion of natural time and social time being sequentially divorced.

As raised at the end of the last section however, does this characterization really reflect Bergson's conception of time, given the qualitatively transitional, co-permeating time states for which he argues? The present-centrism that Merleau-Ponty believes underpins Bergsonian philosophy informs his criticism that Bergson's past-as-pure-memory is a weak degree of the perceived present. We have seen though that Bergson refutes this definition in *Matter and Memory* of "the difference between actual sensations and pure memory as a mere difference in degree, and not in kind" (Bergson 2004, 179). Recent debates on this matter are most prominently symbolized by Leonard Lawlor's argument that the position for which Merleau-Ponty criticizes Bergson is not found in *Matter and Memory*. As a result, an unlikely alliance is supposed between Merleau-Ponty's and Bergson's conceptions of the past-present relation, where Lawlor states that "if Merleau-Ponty rejects the conception that he incorrectly attributes to Bergson, then he actually supports Bergson's position" (Lawlor 2003, 90). In further situating the preceding interaction between Merleau-Ponty and Bergson within subsequent commentaries, we see that because of their mutual impressions of "the interpenetration of past and present" (Cohen 1999, 28), Richard Cohen posits that the "true ally of Bergson is . . . Merleau-Ponty" (1999, 27). While I am in agreement with Cohen regarding the intersection between the two thinkers on certain points, perhaps given the preceding considerations it is a stretch to call them "true allies."

Given such marked disagreement about Bergson's position, inviting a philosopher with a lengthy investment in the Bergsonian project can assist us. For Merleau-Ponty, memorial past and perceptual present *co-constitute simultaneously*. What will be seen, however, is that somewhat similarly to Lawlor, this is also how Gilles Deleuze interprets the Bergsonian past-present relation, contrary to Merleau-Ponty's reading of Bergson. Deleuze will assist us intermittently from this point in this chapter. Just as it would have been remiss to explore philosophies of time without engaging Bergson, similarly it must be observed that Deleuze commands a certain ubiquity in Bergsonian scholarship. In considering any possible incompatibility between the readings by Deleuze and Merleau-Ponty, the earlier point concerning the heuristic nature of this Bergsonian exploration should be recalled. My intention is not to take a "side" in this debate but to use the threads that manifest to consider whether intersecting reflections on philosophies of time can incorporate the divided conditions of social lateness within a monistically synchronous temporality.

In exploring the Bergsonian past-present relation, the duality of the memory-image attracts primary attention. This is due to how Bergson describes it in *Matter and Memory* as simultaneously partaking of both the pure memory, "which it begins to materialize" (Bergson 2004, 170), and of perception, "in which it tends to embody itself" (2004, 170). Perception is duly conceived as "impregnated with memory-images which complete it as they interpret it" (170). Deleuze, in *Cinema 2: The Time Image* (1989 [1985]), interprets this commonality between perception and memory as the primary co-implication, *not* the secondariness of an opposition, of present and past states. While it is necessary for the present to pass (or to become past) for another present to arrive, Deleuze believes Bergson illustrates that it is also "necessary for it to pass *at the same time* that it is present" (Deleuze 1989, 79; my emphasis). The perceptual-memorial memory-image must, in Deleuze's reading of Bergson, be "still present and already past, at once and at the same time" (1989, 79). Deleuze is arguing that this *simultaneous* coexistence conditions time, similar to our argument developed phenomenologically in which time states are simultaneously co-constitutive. If the past did not exist *with* the present, the present would not also qualitatively transition and time would not be.

Such co-constitution means that the Bergsonian past/memory for Deleuze does not reemerge as the present/perception it once was, a subordinated, fixed, distant past that is simply (re-)called. Rather, the past participates, in Deleuze's terms, in "a different present from the one that they have been" (Deleuze 1989, 79). Neither present nor past is hierarchized, meaning we must now be careful with commentaries such as that provided by Lawlor. Lawlor recognizes the co-constitutive states that are also a feature of Merleau-Ponty's phenomenology (Lawlor 2003, 90). Curiously though, Lawlor's terminology could be seen to compromise the primitive implication of the present with/as the originary past, in suggesting that "the present itself is *dependent* on a past, on the 'original or originary past'" (2003, 90; my emphasis). A more nuanced description of this would be to recognize, following Merleau-Ponty, that neither time state is solely "dependent" upon any other.

Nevertheless, both Deleuze's and Lawlor's respective arguments reconfigure Merleau-Ponty's reading that Bergson hierarchizes present time states. If presents were untouchable, solely productive, sources, this would prevent the "radically novel/new" emerging. As Deleuze explains in *Bergsonism* though, such novel/new emergence *is a feature* of Bergsonian time, contradicting the notion of a "real that is ready-made, preformed, pre-existent to itself" (Deleuze 1988, 98). This perspective maintains that for Bergson, time states *are* perpetually, qualitatively novel/new. A consistency duly emerges with the impression that we have reviewed, of how the co-permeating, heterogeneous relations between time(s)-as-duration(s) for Bergson marks their (re-)production as different kinds of pasts and presents, beyond replication of, as Deleuze puts it, "sterile doubles" (1988, 98).

In the previous section, Merleau-Ponty's reading that the Bergsonian past and present are separated/divided/juxtaposed, and that the past is latently powerless, reinforced rather than reconfigured the position that simultaneously divided/juxtaposed states were outside time's real force. In this section we reviewed a different approach, by revisiting Merleau-Ponty's conception of this force as of co-constitutive/co-distinguishable time states and observed that this is actually how Deleuze also interprets the Bergsonian past-present (co-permeating) relation. What therefore eventuates via Deleuze is that again we are unable to accommodate within our Merleau-Pontian model of time from chapter 5, the divided/juxtaposed time states, such as are spatially represented on clocks, which condition the reality of lateness.

It is worth noting though that lateness is not exclusively an "all or nothing" affair in everyday terms. Indeed, there seems to be degrees of synchronization. Someone can be "partially late" if they are perceived to only just miss their allocated meeting time, an act which while not entirely satisfying the requirements of interpersonal synchrony, can somehow be accommodated with such requirements. My interest in this otherwise rather mundane observation is that it bears a certain terminological resemblance to how the co-constitution/co-distinguishability/co-permeability of states that Deleuze has just identified in Bergsonian philosophy is defined by Bergson as a "partial coincidence." Is there a symmetry to be observed between partial lateness and coincidence? In exploring whether Bergson's philosophy of time recognizes not just a global synchrony but also localized and socialized asynchronies, we shall explore this partiality now.

THEORETICAL PARTIALITY: SIMULTANEITY TEMPORALIZES

In considering the relation of Bergson's impression of partially coincident states to the Merleau-Ponty-inspired conception of simultaneously co-constitutive states, let us first attend to Merleau-Ponty's description of such partiality as the *impossibility* of past-present complete coincidence. In *The Visible and the Invisible*, Merleau-Ponty defines a coincidence that is partial as "a coincidence always past or always future" (Merleau-Ponty 1968, 122). This refers to the co-constitution of visibility (the present) and invisibility (the past and future), in which the past (re-)emerges in a new/future form simultaneously with/as the equally novel present. The past is not simply presented again as the present it once was, which is why Merleau-Ponty describes the past as "impossible," evoking the "past that was never present."

This original past is concurrently also an original future given that the novelty of this production of the "past" manifests as the "past" and/as the "future" of the co-constitutive present. To be consistent, I believe these time states are better described as the *past-and/as-present*, and as the *future-and/as-present*, or in Merleau-Ponty's terms, where the present is an "experience that remembers an impossible past, anticipates an impossible future" (Merleau-Ponty 1968, 122–23). We have considered how the co-distinguishability/co-constitution of these states conditions time and/as Being, whereby in (re-)productively emerging from nothing but itself, Being originarily/novelly emerges from itself as the self it *always already was*. I will describe this as "always already *impossibly* was," in keeping with the current theme, where one's inescapably being-present-in-the-world means any recourse to the "was" of Being is always already co-constituted with the experiential present.

Through reconfiguring what we take for granted about "distance," Merleau-Ponty manages the same, seemingly irreconcilably, opposed positions with which we are grappling: divisibility and distinguishability, or transcendence and immanence. His attempted co-accommodation of these concepts describes a monistic subjectivity in which "it is only by remaining at a distance that it remains itself" (Merleau-Ponty 1968, 123). This "distance" will be crucial to relating Bergsonian *partial* coincidence to Merleau-Pontian simultaneity, and in turn to the possibility of accommodating temporal divisibility with temporal co-constitution.

There is distance in coincidence for Merleau-Ponty. However, this distance is an immanence, as Being is said to be *relieved* from itself via forms of itself, "a relief which remain[s] distinct" (Merleau-Ponty 1968, 123). This is a strange form of distance, maintaining an internal self-presence while distinguishing Being-as-self from what would impossibly otherwise be a static, metaphysically transcendent presence. This self-divergent presence is attributable to impossibly present pasts, which in presenting are never present to be recalled as they previously were present. Such self-relation marks the implicated simultaneity of past and/as present, where as Merleau-Ponty eloquently states (and in naturally material terms no less), "the weight of the natural world is already a weight of the past" (1968, 123). Moreover, I suggest the weight of the present is already a weight of all other times. This simultaneity of weighty materialities has in chapter 5 been recognized as a form of social synchronization. If every materialization constitutes, and is constituted by, other weighty materializations, subjects and objects co-temporalize simultaneously. Or in Merleau-Ponty's terms, "the world and I are within one another, and there is no anteriority of the *percipere* to the *percepi*, there is simultaneity" (123; author's original emphasis).

If we revisit the topic of theoretical meta-analysis, I observe that Merleau-Ponty's phenomenological critique does not assume a position of objective externality, examining an established world on which theory can provide wisdom. Rather, in Merleau-Ponty's reflection on the nature of reflection, we see an awareness that his theorizing, perceiving existence, his "I," embodies a theoretical perspective of Being. By perceiving, reflecting upon, and reproducing Being's material conditions, such theory is never blinded by what we have seen Hoy, Horkheimer, and even New Criticisms describe as the established structures of the present. Instead there is an attention on the *percipere*-I as theory, as a material presence that is aware of how it is materialized/presented by the very material/present that it perceptually/theoretically materializes/presents, and that *it is*. This self-awareness of self-germination duly fulfills a criteria by which we have seen that critical theory historically self-defines.

Merleau-Ponty's interpretation of this reversibility and simultaneity coheres with the conception developed of co-constitutive/co-distinguishable/co-permeating time states. It has been earlier argued that the distinguishability of materializing subjects in/as each other opens each subject onto, and into, itself, concurrently (re-)producing its impossibly preexisting self-as-time(s). This (re-)production marks the perpetual originality of the past "*percipere* I," the subjectivity-as-past, and thus the impossibility of identically reliving a previously present subjectivity. Merleau-Ponty describes such productive self-divergence/self-distance as "an openness upon . . . the past[s]" that consequently "enter[s] into their definition" (Merleau-Ponty 1968, 124). In being recollected, the past "as it was" is "inexplicably altered" (1968, 124), perpetually (re-)emerging as a simultaneously co-constituted past-and/as-present, characterizing the impossibility of a past that is simply as it was when it was previously present.

Can we say that Bergson's sense of "partial coincidence" is similarly receptive to the reconfiguration of divergence, distance, and separation? For Bergson, the past manifests as something with which the perceptual present "partially coincides" (Bergson 2004, 292, 297–98). Coincidence is "partial" because the perceptual present is not identical with the past that it will become once it forms relations with new presents. Indeed, there is a prominent aspect to Bergson's argument that characterizes each perceptual present as only manifesting the *utilitarian* aspects of past presents (2004, 206), or what fits about the past "from the point of view of the action to be accomplished" (220). As with the perpetual "inexplicable alteration" of past-and/as-present just encountered via Merleau-Ponty, in this regard Bergson defines the partial coincidence of the present with the past not as of preexisting, divided states but as the creation of states that are "new every moment" (297).

However, one could also interpret here the reiteration of Bergson's earlier encountered separation of socially pragmatic phenomena from that of real phenomena. Consistent with the social constructionist thought that would follow his era, for Bergson a complete, true reality is recreated by *novel*, variable, partially incomplete present versions that represent it according to what "fits." If we recall Bergson's conception of intuition, it is as a *method* that intuition opens awareness to how states of time are not simply representations of quantitatively different, spatial degrees but are also more fundamentally, qualitatively different, co-permeating states. This co-permeation of "partially coincident" pasts and presents is where, for Bergson, the present as "concrete extended *recovers* its natural continuity and indivisibility" (Bergson 2004, 293; my emphasis). Partial coincidence here thus describes a mode for Bergson that comprises a discontinuous, perceptual, spatial, socially pragmatic present that is *rediscovering* its co-permeating, indivisible, durational/memorial truth. This seems to be a *return* to primordial or real (the social constructionist's "natural") conditions, where in Bergson's terms, the perceptual present and the memorial past are "grafted upon the other" (2004, 297). Indeed, Bergson describes partial coincidence as the radical form by which, for matter and memory or perceptual present and memorial past, "*union* becomes possible" (297; my emphasis).

It is seemingly according to such logic that Merleau-Ponty demands in *The Visible and the Invisible* that Bergson's conception cannot recognize a singular ontologizing reality, but rather situates real time outside the present, social utilization of it. This is due to Merleau-Ponty's interpretation that Bergsonian partial coincidence attempts to "return to the immediate data"

(Merleau-Ponty 1968, 124). Bergson's use of terms such as "recovers" and "union" likely informs this contestation from Merleau-Ponty to the partial reconciliation of past-present coincidence. This is an intervention that is consistent with Merleau-Ponty's general critique of "intuitionist philosophies," which posit, or anticipate, a return to the natural/inherent/real immediacy of worldly givenness. For Merleau-Ponty, and for our position developed in the last chapter, the perceptual present is co-constituted with/as the memorial past. However, the originary simultaneity of past and/as present that we have comprehended does not, in Merleau-Ponty's terms, suggest their union, fusion, or coincidence (1968, 123, 124, 127, 128, 191). Bergson conceives of the relation between the perceptual present and memorial past as where the perceptual present partially coincides with, unifies with, or recovers its own "indivisible continuity." This characterizes partiality as a distance/dehiscence that, according to Merleau-Ponty, can only be a "bad or abortive truth" to be overcome (125).

Conversely, we have posited that Being's chiasmic distance/divergence/distinguishability actually temporalizes its emergence and self-consciousness. This is attributable to the Merleau-Pontian inspired appreciation of the ambiguity of temporalizing worlding subjects (Merleau-Ponty 1968, 117, 118, 123, 128, 145, 146, 153). What Bergson characterizes as "only/badly/abortively" partial is instead described by Merleau-Ponty as a "good error" (1968, 125), not preventing an entire coincidence with Being that a subsequent "union" only partially addresses, but perpetually "opening" Being's being. In supposing that Bergson negatively characterizes partiality as incompletion, Merleau-Ponty argues that this neglects how "every being presents itself at a distance, which does not prevent us from knowing it" but is "on the contrary the guarantee for knowing it" (127). Being distinguishes itself from itself as concurrently co-conditioning subjects/knowers and objects/knowns. From this insight I have redefined social constructionist thought from a set of inquiries restricted to critiquing humanity's partial representation of a preexisting, inaccessible worldly reality, to the world's immanently self-representational reality.

Merleau-Ponty is arguing that when Bergson uses the term "partial" he refers to something preexisting, hidden inaccessibly in-itself. What "Bergson lacks" here, according to Merleau-Ponty, is to recognize "the identity of the retiring into oneself with the leaving of oneself, of the lived through with the distance" (Merleau-Ponty 1968, 124). Partial coincidence is not something that prevents the present's complete coincidence with an aspect of the past and memory by imposing a distance-as-void. Rather, this distance, this partiality, conditions the complete possibility of present, past, and memory, via/as the simultaneous co-distinguishability of self from/as self by which Being temporalizes. Merleau-Ponty duly advises Bergson that partial coincidence is "a Self-presence that is not an *absence from oneself*," but instead is "a contact with Self *through* the divergence *(écart)* with regard to Self" (1968, 192; author's original emphasis).

This temporalizing/ontologizing quality of partial coincidence/simultaneity is what Bergson neglects according to Merleau-Ponty, evidenced in language and forgetting. We will first briefly discuss language. Earlier it was exhibited that Bergson describes language as an intellectual (mis-)representation of an inaccessible reality, evocative of the polarizing, social constructionist time condition that posits human productions such as social time structures as mediated representations of a separately natural, real-world temporality. Language for Bergson (particularly scientific and philosophical discourse), spatializes, homogenizes, and separates from intuitional processes what consciousness would otherwise experience directly. This is the point that Hulme's poetry theory has been seen to adopt, informing aspects of New Criticism accordingly. In response, Merleau-Ponty describes in *The Visible and the Invisible* that for Bergson "the philosopher speaks, but this is a weakness . . . he should keep silent, coincide in silence, and rejoin in Being a philosophy that is there ready-made" (Merleau-Ponty 1968, 125). Merleau-Ponty's concerns here about assumptions of a transcendent state of Being evoke his previous reconfiguration of the phenomenological method.

While Bergson characterizes language as the socialized human's intellectual break with the intuitive flow of consciousness, for Merleau-Ponty language describes the chiasmic/simultaneous way in which Being becomes, self-consciously. This is a continuation of earlier arguments that theory, critical or otherwise, is Being's self-analysis and self-expression rather than a retrospective reflection from a culturally constructed, external position. Language incarnates as Being's distinguishability, whereby in Merleau-Ponty's terms, the "dehiscence of speaking" is what "opens it [Being] to itself" (Merleau-Ponty 1968, 117). Conversely, semantic philosophies such as Bergson's are said to reductively "close up language as if it spoke only of itself" (1968, 126). While I agree with Merleau-Ponty's criticism against conceiving of language as only speaking of itself *as an exclusively human linguistics*, I would argue that language *does* only speak of itself. This requires the interpretation that language, as theory, as philosophy, is Being itself and Being's life, or as Merleau-Ponty himself puts it, "language is a life, is our life and the life of the things" (125). All incarnations/things by which Being emerges from itself, as itself, are the language of Being. Being duly speaks only (of) itself.

Merleau-Ponty's characterization of language as Being-incarnation, rather than as a representational break, informs a consequent critique of Bergson's portrayal of forgetting as an "occultation" (Merleau-Ponty 1968, 194). Forgetting is an interesting topic in terms of considerations of the social reality of lateness, given that often forgetting is literally a cause or condition of lateness. But is one's forgetting to be somewhere, or to meet someone, or to do something, as straightforward a matter as an indication of something "lacking" in collective consciousness or worldly synchrony?

Merleau-Ponty's critique of Bergson's characterization of forgetting as an "occultation" refers to Bergson's conception of any state apart from the conscious present as lifeless, "latent and unconscious" (Bergson 2004, 181). If during the passage through the present, "a segment of the past would fall into oblivion" (Merleau-Ponty 1968, 194), as Merleau-Ponty identifies in the past-present relation posited by Bergson in *Matter and Memory* (Bergson 2004, 224, 319–20), then aspects of the past would be lost. Conversely, Merleau-Ponty's point is that forgetting is not an inability to access (remember) a purely past present. This is because an already originated, established, pure in-itself past, never manifests.

Extending Merleau-Ponty's thesis, I would argue the past present that is "forgotten" is actually the "impossible past," the past that was never simply present, as it is always becoming a co-constituted present-and/as-past. This past does not "latently" remain to be remembered, given that "forgetting" this pure past (a past that never occurs) actually opens the past to (re-)articulation with/as the present. This is a somewhat Bergsonian point on intensive states, in that for the past to (re-)emerge in and with the present, it must be *different in kind* to what it was. Nevertheless, Merleau-Ponty criticizes Bergson's converse discussion of the past as preserved "pure memory," stating that "if the pure memory is the former present preserved . . . it becomes impossible to see how it could open to me the dimension of the past" (Merleau-Ponty 1968, 122).

For Merleau-Ponty, partiality, forgetting, is the originary condition of past, present, and ironically, memory. Where Being relaxes-and-contracts, forgets-and-remembers, or perceives-and-is-perceived-by itself, within/as itself, incarnation manifests. This forgetting that conditions self-remembering is mimicked by Merleau-Ponty's correlative argument that imperception is inherent to perception (Merleau-Ponty 1968, 225, 247). Equally, this recognizes the invisible constitution of vision, as the impossible past that constitutes the present, or the constitutive "unconscious of consciousness" (1968, 255). Being-consciousness, the materializing of time, perpetually finds itself by forgetting itself. To respond directly, therefore, to a lingering question regarding the extent to which the Merleau-Pontian perceptual present is material, such insights determine that it is the immateriality of materiality that defines, for Merleau-Ponty, perceptual materiality. I would argue here that theory, as materiality, perceptually distances itself representationally from materiality, while it perceives materiality, *as* materiality. This distancing is a materializing, where what is material is not simply the "naturally real world" from which socially constructed theories are subsequently divergent but a reality that is a socially constructed forgetting/remembering of such reality.

In arguing that Bergson misses his own point regarding past-present co-implication, Merleau-Ponty exhibits traces of a deconstructive method in his intervention. It is according to such logic that Jack Reynolds argues for the similarity between Derridian and Merleau-Pontian methods, "notwithstanding Derrida's efforts to distance himself from this tradition" (Reynolds 2004, 82).[9] Reynolds reminds us of the general scarcity of commentary on Merleau-Ponty by Derrida (2004, 61). However, Derrida's only sustained dialogue with Merleau-Ponty, in *Memoirs of the Blind* (Derrida 1993), is "predominantly flattering" (Reynolds 2004, 61). Merleau-Ponty's argument concerning the imperceptive constitution of perception, or the invisibility that is inherent to visibility, is actually affirmed by Derrida who states that "invisibility would still inhabit the visible . . . the visible as such would be invisible, not as visibility . . . but *as* the singular body of the visible itself" (Derrida 1993, 51; author's original emphasis). Invisibility is not the impotence of visibility for Derrida, nor for Merleau-Ponty. Rather, invisibility-forgetting is the condition of visibility-remembering (and vice versa).

If apparent binary opposites are in fact co-constitutive, how might this inform our attempt to unsettle the apparent opposition of simultaneously divided/juxtaposed, versus simultaneously co-constitutive/co-distinguishable, temporalities? Merleau-Ponty's critique is that partially coincident states are simultaneously co-constitutive states, where partiality is the necessary internal distance from which any temporal state or temporalizing manifests. This is contrary to what he argues is Bergson's restricted sense of partiality, as the never-quite-achieved traverse of a division/abyss *between* time-states. Having reviewed partial coincidence we cannot shift Bergson's seemingly entrenched position that simultaneous states are spatially divided/juxtaposed outside time's co-permeating, durational, real conditions.

Merleau-Ponty further demands, in response to Bergson's ongoing exclusion of time states that exhibit spatial division, that it is "neither necessary nor sufficient to condemn the spatialization of time as does Bergson" (Merleau-Ponty 1962, 482). Perhaps, in developing Merleau-Ponty's just encountered interpretation that the partial co-constitution of states is the opening of all temporalizing, we might find that spatially divided times, and non-spatial, co-permeating/co-distinguishable times, are not opposed for Bergson?

MONISM IS DUALISM, SYNCHRONIZATION IS LATENESS

Merleau-Ponty's accusation that Bergson recklessly condemns the spatialization of time must first be qualified. We have seen that for Bergson, the lived experience of time is constituted by; (1) spatial, measurable, quantifiable discontinuities, the necessities of social pragmatics, *and by* (2) durational, continuous, qualitatively co-permeable, intensive states. Both temporalities are experiential realities. Consequently, we are dealing with a more nuanced argument than a straightforward condemnation of time's spatialization. Instead, Bergson's conviction concerns the assumption that spatialized time accounts for *all* that time is.

Merleau-Ponty argues that a spatial conception of time is not a problem in terms of time's reality if our primordial being-in-the-world is acknowledged. This being, a materializing, spatializing, fleshing-as-temporalizing where we never finalize as a material space/flesh, means that any "time is exclusive of space only if we consider space as objectified in advance, and ignore that primordial spatiality . . . which is the abstract form of our presence in the world" (Merleau-Ponty 1962, 482). Bergson's stigmatization of the spatial constitution of time is accordingly not a sufficient basis, in Merleau-Ponty's estimation, to provide an "authentic intuition of time" (1962, 482).

I agree with Merleau-Ponty that Bergson's exclusion of spatial time is not an adequate method to define "real, intuitive time." My reasoning can be linked to Bergson's demand in *Matter and Memory* that "each unique moment of the past survives" (Bergson 2004, 179). What does this mean, that the "past survives"? Rudimentarily, this is something with which our considerations would cohere. If no state of time exists in-itself, but (re-)manifests with all other states of time, then every time is always somewhat "existing/surviving." But why, we should ask, is it specifically important to Bergson that the past survives?

To answer this, we must note that for Bergson, if something spatial is not perceivable, if it is outside our field of vision, it can still be assumed to exist. This coheres with his conception of spatial simultaneity, in which spatial things are simultaneously dispersed in "space, thus appearing to preserve indefinitely the *things* which are there juxtaposed" (Bergson 2004, 184; author's original emphasis). With unperceived ideal states such as the past, however, this preservation is not always granted. Past states are assumed to have once existed, but only before the passing of time, whereby "time in its advance devours the *states* which succeed each other within it" (2004, 184; author's original emphasis). Bergson asks why the existence of an unperceived spatial state is accepted whereas the existence of an unperceived ideal state is not. He demands in response, in referring to the involvement of ideal/memorial states in perceptual/present states, that "the adherence of this memory to our present condition is exactly comparable to the adherence of unperceived objects to those objects which we perceive" (187).[10]

Having therefore refuted the difference between the existential state of spatial *and* ideal objects that are not immediately perceived, Bergson then overcompensates, in my evaluation, by installing a new difference. He does this by asserting that ideal objects that are not being perceived are *always lived*. This, he posits, *cannot* be said of spatial objects outside our perceptual field. The co-permeation of ideal/conscious states means that "our previous psychical life exists for us *even more than the external world*, of which we never perceive more than a very small part" (Bergson 2004, 188; my emphasis). While this adheres to Bergson's established distinction between intensive and extensive states, a problem I identify is that when installing this new difference, Bergson still recognizes that there are infinite, unperceived links between material or spatial objects being perceived and material or spatial objects not being perceived, since all "obey necessary laws" (2004, 190). Even though unperceived space is linked to perceived space as that which "hide[s] behind it infinitely more than it allows to be seen" (190), it is denied that we live these unperceived but connected spatial states. Contrarily, Bergson affords no such hiding or experiential exclusion to the ideal/intensive states comprising time-as-consciousness. Given that intensive/conscious states are not juxtaposed like extensive/spatial states, a co-permeating co-constitution is posited for the ideal realm in which "our past psychical life conditions our present state" (191). Consistent with Augustine and Durkheim, and contrary to Merleau-Ponty's interpretation of his position, for Bergson memories are not stored somewhere. As Bergson asks about the past, "[I]f it is retained, where is it?" (191). For Bergson, such a question concerning *where* the past or memory is relies on "images drawn from space" (191). Instead, time/ideal/conscious states are said to co-disperse, refuting notions of finite, spatial locations.

Despite this distinction, what I want to consider is whether Bergson's claim that each moment of the past exists for us, but each aspect of our spatial world does not, can be supported according to his logic. If we review Bergson's reasoning, his first move correlates the existence of "not presently perceived" spatial/extensive/material objects/states with "not presently perceived" time/ideal/conscious objects/states. This is unusual, given the manner in which Bergson usually demands mutually exclusive properties between time and space. Nevertheless, spatial/material states not being directly perceived still presently exist, and there is no reason, Bergson tells us, that not directly perceived time/ideal states such as the past should not also presently exist.

Given that a premise about space leads Bergson to a conclusion about consciousness-as-time, is the consequent distinction he draws between the two regarding their respective ongoing existences valid? By (a) using an argument concerning the coexistence of space(s) in order to explain the coexistence of time/ideal/conscious states, and then by (b) justifying the coexistence of time/ideal/conscious states through their co-constitutive implication in/as each other, should it duly follow that the coexisting spaces from point (a)—the original premise of the logic about time/ideal/conscious states—exhibit the same co-constitutive implications as time/ideal/conscious states are attributed in point (b)? This also seems possible, let us not forget, given that Bergson concedes infinite, unperceived links, between spaces.

Bergson of course conversely argues that the time/ideal/conscious states of our previous (not present) psychical life exist "more" than the spatial/extensive states that are not currently perceived (not present). However, as per the above breakdown, in recognizing time/conscious/ideal states as co-constitutive, I am inclined to believe that space should also have these characteristics. This re-reading of Bergsonian spaces would describe not simply a quantitative multiplicity. Rather, it would also conceive of such spaces as qualitatively transitional, and perpetually co-constitutive/co-distinguishable/co-permeating. Or in terms of the social constructionist time question specifically, we would be asserting that spatially divided time states, and chapter 5's co-distinguishable time states, are consubstantially embedded in a singular reality of time.

This dual character is in fact indicated in *Time and Free Will*, where Bergson surprisingly concedes that a quantitative multiplicity is qualitatively conditioned. What is spatial and quantitative cannot be ascertained "without considering at the same time a qualitative multiplicity" (Bergson 1960, 122–23). The argument is that by quantifying/counting objects, which are assumed to be discontinuous, juxtaposed, and independent of their being quantified, there are also qualitatively different relations that manifest between such objects. The process of quantification becomes a constitutive aspect of the things being quantified, and therefore also of the quantified/unified whole. As Bergson states, quantifying them "alters the nature, the appearance and, as it were, the rhythm of the whole" (1960, 123). Bergson goes on to insist that the spatially discontinuous representation of homogenous time, that "numerical multiplicity" of "well-defined states" (128), is actually constituted by the conditions "below" it, "a qualitative multiplicity" whose "heterogeneous moments permeate one another" (128). It was upon encountering these passages that I sensed an invitation to interrogate Bergson's supposed exclusion of heterogeneous time-as-duration from homogenous, spatial time.

Is this sentiment also not true of our experiences with spatialized, symbolized, social time? In quantifying time as seconds, minutes, hours, days, and so on, do we not create, or "alter," the rhythm of quantified time? This I believe was seemingly Durkheim's point, that the regularity of time is inextricably bound up in the social construction of time concepts because it is through divisibly quantifiable concepts that time is invested with shared or collective meaning. When considering how lateness occurs, therefore, due to the variable way in which socially quantified time is lived, it could be said that the qualitative differentiation of a quantitative phenomenon emerges in how differing calculations and navigations of time's collectively symbolic form affect its objective value between individuals.

The simplicity of this reading of Bergson would potentially help avoid counterintuitive exclusions. How could the quantification of spaces not produce, one might ask, a qualitatively different relation between such spaces, in which each is simultaneously, separately identifiable from the others and yet also constitutively implicated in/as the others? How could this not produce quantified-pasts and being-quantified-presents that co-transition into different pasts-as-presents? What is qualitatively relational, and co-constitutive/co-distinguishable, conditions what *seems to be only* quantitative and juxtaposed/divided. Bergson would apparently agree, given his exclamation that "it is through the quality of quantity that we form the idea of quantity without quality" (Bergson 1960, 123). If qualitative and quantitative states instead share immanent conditions, then divided time, the time of homogenous, juxtaposed, points on clocks and calendars, can be accommodated within a monistic temporality of co-distinguishable time states, as can be the social reality of lateness that such juxtaposition conditions. Monistic synchronization, and localized, socialized asynchronies of lateness can seemingly be exhibited within Bergson's considerations about coexisting temporal realities.

Simultaneously, spatially divided states, are the condition of social lateness. Whether via the mechanical representation of clocks and calendars, or the differential positioning of the sun in the sky, a conceptually divided space regulates the common agreements by which humans socially synchronize as well as adjudicates any contradiction to those agreements when someone is late. By engaging Bergson's theory of time, with some considerable critical assistance from Merleau-Ponty, what we see is how one might suggest that divided states do not socially, representationally function *outside* a global or monistic synchrony.

Deleuze quite contextually for our current discussion affirms the perspective that Bergson's philosophy exhibits these monist, *as well as* dualist, tendencies (Deleuze 1988, 73).[11] My reading of Deleuze can be situated among recent commentaries[12] in a manner which interprets that according to Deleuze, in Bergson's work there are the dualisms of "pure present and pure past, pure perception and pure recollection" (1988, 74). However, present/perception also monistically co-constitutes past/recollection, where "the present is only the most contracted level of the past" (74). The inverted base of the Bergsonian cone represents the dispersed past-as-memory for Deleuze. As the cone *contracts*, it reaches its point. This point is the perceptual present, which as seen in *Matter and Memory* is constituted by relevant past states, "an incalculable multitude of remembered elements" (Bergson 2004, 194). It is this present-as-past-contraction that indicates a monism for Deleuze, where past and present only differ according to their relatively relaxed or contracted states. Or in Deleuze's terms, they "have only differences of expansion and contraction" marking "an ontological unity" (Deleuze 1988, 74).

If past and present differ only in terms of expansion and contraction, then their common ontological constitution means they must differ qualitatively/heterogeneously to avoid blurring into each other indistinguishably. That they are distinguishable is perhaps what indicates for Deleuze, echoing Bergson, how "quality emerges from this, quality that is nothing other than contracted quantity" (Deleuze 1988, 74). The interpretation I adopt here is that dualist division/juxtaposition is accommodated with monist co-distinguishability/co-constitution via the *impossibility* of the dualist/divided past. This is the earlier encountered, Merleau-Pontian impossible past that was never present. The impossible past will always be divided/juxtaposed/discontinuous/separated from its dualist pure present because it never was, and never will be, simply a past/previous present. The impossible past does present, it does become a present, however, that is only by (re-)manifesting with/as a present that it co-conditions: the monistic past-as-present.

Indeed, to be consistent with our Derridean engagement concerning originary all-co-constitution, we should describe this as the impossibility of ontological divisibility. Impossibility primordially co-conditions possibility, neither preceding the other. Merleau-Ponty has also alerted us to forgetting being the immanent origination of memory. It is this kind of concurrent divisibility/impossibility and co-distinguishability/commonality of Being that in Deleuze's terms "allows us to go beyond the duality of homogenous quantity and heterogeneous quality, and to pass from one to the other in a continuous movement" (Deleuze 1988, 74). I can readily apply this logic to this book's theme: lateness. The forgetting of being on time is the remembering of lateness, and vice versa, whereby forgetting or lateness do not represent the social's momentary abyss but its monistic-dualistic production. Barbara Adam's concern in chapter 5 about entirely "out of sync" states of Being (humans and CFCs), which I characterized as problematically requiring a temporal abyss, can here be recontextualized. Such states are now defined as "forgetting" their synchrony, from which a dualist relation manifests. However, according to the reading developed, this forgetting is still a mutual production, from which the realities of synchrony and asynchrony are simultaneously and co-constitutively conditioned/remembered.

The perceptual/mattered present, and the memorial/durational past, are monistically constituted according to the Deleuzian logic, for what is extended if not that which is contracted, and vice versa? This is why Deleuze concludes of Bergson that "there is always extensity in our duration, and always duration in matter" (Deleuze 1988, 87). It seems preferable to say duration *is* matter, and vice versa, rather than follow Deleuze's Bergsonian terminology of extensity "in" duration. We have seen Merleau-Ponty criticize this type of terminology as planting the "seeds" of one realm in the other, separating the constituents of what should always already be an ontological unity. Deleuze himself indeed argues against conceiving of a mixture of duration and matter, as though we "begin with a composite . . . the space-time mixture" (1988, 95).

As outlined in the introduction, this term "space," in the argument as I have developed it, refers to all materiality/physicality/extension. This is not the case for Deleuze's reading of Bergson. For Deleuze, *matter* is extension and is durational. Conversely, Deleuzian *space* represents matter-as-duration's ultimate, static expansion, as duration's duration-less externality of juxtaposed states (Deleuze 1988, 87). Nevertheless, just as it has been seen that the impossible past is the past that is never simply relived as it was when previously present (because the past is always becoming a co-constituted past-and/as-present), equally I argue that "space" for Deleuze is an *impossible present*. My argument is that the present never exists in-itself as pure, inert space, but only becomes present via a perpetual (re-)co-constitution with past-as-duration, as a mattered present-and/as-past. Deleuze helps here by noting that "matter is never expanded enough . . . to stop having this minimum of contraction . . . through which it is a part of duration" (1988, 88). The tip of Bergson's cone is the perceptual, spatial, extended present as the most contracted point of duration-as-past(s). However, what this point represents is the opening or possibilizing of impossible-space as possible-time-matter, a perpetual implication in/as what is less contracted and dispersed durationally/memorially. In correlative terms I have argued that the contraction of synchronization is implicated in the relaxation of lateness, and vice versa. It occurs to me now that had this book's inquiry into lateness been Deleuzian driven, an apt title would have been *Naturally Late: The Relaxation of Synchronization.*

At the outset of this chapter, I described why a philosophy's reflection upon, and disruption of, separations of primordially monistic synchronization (co-distinguishability), and contingently localized, socialized asynchronies of lateness (dualist divisibility), would indicate a critical capacity. Whether we can attribute Bergson-the-theorist with this kind of criticality is contentious, given that he would probably be reluctant to agree with the kind of conclusion that this intervention attributes to his argument. This is despite it being a conclusion that has been developed via the terms of his arguments spanning many texts.

Perhaps more important in any case are the critical cross-pollinations enacted between, and reflected upon by, different philosophical methods. Derrida's deconstructive critique began this examination of philosophy's capacity to reconfigure the social constructionist time condition, with the traces of its method also being identified as apparent in the chapters that followed. Merleau-Ponty's phenomenological reflections on, and recalibrations of, phenomenological methods took our inquiry in a monistic direction but could not itself, according to my reading, reach the conclusion at which we have arrived in this chapter. What was required was the complement of the Bergsonian perspective, which when presented in tandem with Merleau-Ponty's critique of it affirmed that theorizing is not only a separately subsequent representation of, and contestation to, a position that precedes but is also the materialization and production of a present position. In studying secondariness as lateness, we have been moving within these kinds of interpersonal, intertheoretical, past-present synchronies. However, within such synchronies is a reality of lateness that has been maintained as primarily internal rather than secondarily external. As the timings of times have been appreciated accordingly, we might have reflected on any presumptions we hold, that there is a pure, externally prior, naturally real state of time that was not always already itself a kind of representing.

Notes

INTRODUCTION

1. As can be seen from the immediately preceding discussion, I use the terms "space" and "matter" interchangeably in order to direct us toward a sense of their substantiality. This interchangeability will be examined in the latter stages of the final chapter's engagement with the philosophies of Gilles Deleuze and Henri Bergson, in which a distinction between spatiality and materiality will be considered.

2. Whether the term "arrive late" is a strictly grammatically correct construction is a valid question. An adverb could be more exact here than the adjective "late," perhaps "tardily," given that "lately" has other meanings which might confuse matters. However, given the sense that I am trying to portray in this book regarding the specific term "late," complementing "arrive" with an adverb like "tardily" seems compromised. Consequently, I defer to the power of common parlance, in which "arrive late" is used regularly, is widely comprehended, and conveys the sense of arriving in an untimely manner that I wish to engender.

1. SOCIAL TIMES

1. For a similar reading, see James Bohman's commentary regarding how Horkheimer's definition is pivotal for the Frankfurt School, in that it defines its approach to critical theory not merely as having "a specific end to be achieved, but rather identifies a distinctly practical activity" (Bohman 2003, 92).

2. This is an approach that is not entirely unrelated to Max Weber's attack on the "overrationalization" of modern society, in particular within academic and scientific institutions, as expressed in the 1919 lecture and publication *Science as a Vocation* (later published as *The Vocation Lectures* [2004]).

3. One example of such an interpretation is found in Stephen Bronner and Douglas Kellner's "Introduction" in *Critical Theory and Society* (1989).

4. Papert's position on the learning of scientific knowledge is evidently informed by Piaget's assertion in *Genetic Epistemology* (1970); "[S]cientific knowledge is in perpetual evolution; it finds itself changed from one day to the next. As a result, we cannot say that on the one hand there is the history of knowledge, and on the other its current state today, as if its current

state were somehow definitive or even stable. . . . Scientific thought, then, is not momentary[;] . . . it is process" (1970, 2).

5. Psychologist Lev Vygotsky's *Thought and Language* (1986 [1962]) is one such example of further, related theory. Vygotsky applies Piaget's theory of cognitive maturation to his own considerations about how children construct knowledge when learning alone versus when under adult supervision. Similarly, George Kelly's "personal construct theory" (1991) is of relevance, in describing individuals as "naïve scientists." This posits that individuals anticipate or predict events, construct psychological processes through those anticipations or predictions, and enact such constructs through the actions they impel. The validity of the construct is determined by how the results of the impelled actions cohere with what was anticipated.

6. For a more expansive discussion of this point, see how Manuel Arriaga articulates the differences between critical theory and traditional theory, the latter of which he argues has "tended to claim to be transcendent to its social basis" (Arriaga 2006, 39).

7. See Kwame Anthony Appiah (1996) and Naomi Zack (2002). Zack's "race elimination" theory, for instance, argues that "there is no coherent explanation of what makes one population, such as inhabitants of sub-Saharan Africa, a race, while another breeding group, such as Protestants in Ireland, would fail to be considered a race" (Zack 2002, 69).

8. "The category of "race naturalism" has many versions. Robin Andreason defends the "cladistic" classification, which defines races as "ancestor-dependent sequences of breeding populations that share a common origin" (2004, 425). Conversely, Massimo Pigliucci and Jonathan Kaplan present the naturalist's "ecotype" characterization of race (Pigliucci and Kaplan 2003) as referring to how geographically distinct populations have evolved a specific method of adapting to their local environment. For Neven Sesardic, the question of human race should mirror the distinctions made between nonhuman animals, noting that "in biology, the concept of 'race' is often regarded as synonymous with 'subspecies'" (Sesardic 2010, 48). Somewhat relatedly, the belief in recognizing race as a natural category is presented from an epidemiological perspective by Neil Risch, Esteban Burchard, Elad Ziv, and Hua Tang (2002), who challenge the argument that race is not coupled with human genes. Their work informs a later article in *New England Journal of Medicine* (Burchard et al. 2003), in which recognizing race as a natural category is demanded in order to better manage public policies regarding health differences between racial populations.

9. Karen Barad similarly questions the notion of an objective physical reality outside socially produced technologies by which such physical reality is apprehended. This reconceives the "physical, natural arena" as a production that occurs in tandem with scientific subjectivities. For a fascinating example of this, see Barad's discussion of how Otto Stern and Walther Gerlach's experimental results, when demonstrating "space quantization," embody a scientist's cigar breath and social status (Barad 2007, 162–67).

10. Such a debate can be viewed as the latest instalment (at the time) of the "two cultures problem," famously described in a 1959 Rede Lecture by Charles Percy Snow as the ideological polarization of scholars in the humanities and associated disciplines, from those in the natural sciences (Snow 1998 [1959], 14–15). Implicit to this division is the assumption that the natural sciences are concerned with discovering singularly objective facts about the world, whereas the humanities (and the social sciences) produce contingently speculative theses.

11. In *The Mathematics of the Calendar* (2007), historian Marc Cohn explores many such historical influences into the Western, seven-day week.

12. For one influential example, see Eviatar Zerubavel's *The Seven Day Circle: The History and Meaning of the Week* (1989).

2. RELATIVELY LATE

1. In *On Time, Punctuality, and Discipline in Early Modern Calvinism* (2009), historians Max Engammare and Karin Maag explore how sixteenth-century Protestant reformers installed a moral code that prioritizes "punctuality." Also relevant is Foucault's *Discipline and Punish*

(1977 [1975]), which observes the eighteenth-century regulation of individual and social bodies via "a whole micro-penalty of time" (Foucault 1977, 178) for misdemeanors such as tardiness.

2. Further elaborations of this interpretation of the "present-centrism" of time in many African contexts can be found in the research of Keith Snedegar (2007, 27), as well as of Osita Nnajiofor (2016, 253).

3. Also acknowledged by David Landes (1983), Eviatar Zerubavel (1989), and again, Helga Nowotny (1975).

4. This is akin to when Alan Sokal asks "post-modern critics," who he believes are straightforwardly positing that the physical laws of nature are social conventions, to test gravity by jumping out of his twenty-first floor apartment's window (Sokal 1996b, 2). The assertion being; physical reality has a factuality that does not depend on what we think of it.

5. An account of this regulation is found in *The Oxford Encyclopedia of Ancient Egypt: P–Z: Volume 3* (2001), edited by Donald Redford.

6. Now named Victoria Pitts-Taylor.

7. In Richard Geldard's *Remembering Heraclitus* (2000), the relation of change to identity emerges via Heraclitus's question as to whether one is able to step into the "same river" twice, given its perpetual change.

8. So rare that Professor Ted Brown from the New York State Institute for Basic Research reports that progeria has "a reported birth incidence of about one in eight million" (Brown 1992, 1222S).

9. The historical origins of this usage of "late" are unclear, something Andrew Stark explores in *Consolations of Mortality* (2016). In his interpretation of how we contemporarily think of the "lately dead as late" (Stark 2016, 27), Stark notes that this suggests that late is "a synonym not for 'tardy' but for 'recent'" (2016, 27). However, the fact that late also describes a person who is "not arriving," something that can now be said of those who are dead, infers that these two meanings of late "infuse one another in a kind of double image" (27).

10. The origination of the term "body project" is often attributed to Chris Shilling's *The Body and Social Theory* (1993).

11. The Aristotelian categories place every object of apprehension into one of ten categories: substance, quantity, quality, relation, place, time, position, state, action, and affection. Neoplatonist, Porphyry, collates these thoughts in Aristotle's *The Organon* (2010 [circa 50 B.C.E.]).

12. References to the objectivity of social facts are numerous in this text, editor George Catlin noting Durkheim's "stress upon the objectivity to the individual of social phenomena" (Durkheim 1938, xxviii).

3. SUBJECTIVE TIMES

1. We can develop a comparison here with the philosophy of the Stoic, Chrysippus, whose conception of continuous time, contrary to that of Aristotle, holds that no aspect of it is "wholly present" (Stobaeus quoted in Long and Sedley 1987, 304). This is because, for Chrysippus, each part of time is able to be divided into infinitely smaller parts (1987, 304). Such parts of time contain infinite divisions of the past, as well as of the future. The result, as Plutarch explains in his *On Common Conceptions*, is that in the Chrysippean view "whatever one thinks one has grasped and is considering is present is in part future and in part past" (Plutarch quoted in Long and Sedley 1987, 304).

2. In the chapter "*Ousia* and *Grammè*: Note on a Note from *Being and Time*" from *Margins of Philosophy* (1982 [1972]), Derrida builds on Heidegger's critique of the Aristotelian present now, by arguing that time cannot be reduced to a now that is isolated and present (Derrida 1982, 61). A different reading is possible of Aristotle according to Derrida, in which the now is both a determining limitation of time as well as time's "accident" (1982, 61). The Derridian perspective argues in this regard that what is seemingly absent to the present is in fact simultaneously participating in, and as, its "presenting." This deconstruction of the pres-

ence|absence binary symbolizes the Derridian concept-that-is-not-a-concept of *différance* (20), which we will more comprehensively examine in the next chapter.

3. There is a certain awkwardness in relatively seamlessly following a comment about Derridian deconstruction with a discussion on Habermas, given the often turbulent, philosophical relationship, the two initially shared. The critique that Habermas provides of Derridian deconstruction, as well as the eventual mutual respect that they express for each other's work, will be explored in the next chapter, given the primary concern there with the Derridian style of deconstructive critique.

4. The text that Habermas cites here is taken from Daniel Riesman's *The Lonely Crowd: A Study of the Changing American Character* (1950). The same theme is later expanded most emphatically in the first volume of *The Theory of Communicative Action* (1984 [1981]), in which Habermas criticizes the intervention of economic and bureaucratic rationalization into the cultural and social lives of individuals.

5. Our engagement with Halbwachs's conception of collective memory looks at the collaborative ways in which any individual memory is produced and recollected. This is distinguishable from the research of Norman Brown (1990), whose social experiments and consequent publications often discuss the *aggregation* of collective memories, examining how population groups have the capacity to remember more voluminously than individuals.

6. See Nick Crossley's "Sociology and the Body" (2005) discussion concerning how this informs the understanding that Durkheim's sociology is potentially not reducible to a mind|body dualism.

4. (DE)CONSTRUCTED BODIES

1. That these are the principal architects of New Criticism is identified in commentaries spanning many decades, as evidenced in Thompson (1971), Lentricchia (1980), Jancovich (1993), Burt and Lewin (2003), and Baldick (2014). Ewa Thompson indeed notes that it has become "customary to refer to them as to a kind of entity because they represent something next to an agreement on matters of literature, and because of personal connections" (Thompson 1971, 46).

2. While this aspect of the critical theory versus traditional theory debate has been covered in chapter 1, it is worth noting that not all commentators endorse critical theory's characterization of itself in this regard. Colleen Flewelling's interrogation (2005, 62–72) of Horkheimer's claims that critical theory is more future focused provides one such competing voice.

3. This is a point that has been relatively recently explored by Stephen Cohen (2002, 19–21), who likewise raises the ahistorical aspects of New Criticism as being differentiated from the Frankfurt School mandates.

4. Saussure's semiological theory was delivered in a "General Linguistics" lecture series between 1907 and 1911 at the University of Geneva. Students organized and published this material after Saussure's death as *Course in General Linguistics* (Saussure 1966 [1916]).

5. This theme is also read through Derrida's impression of God being positioned as a transcendental signified, as discussed by Collins (2015, 55–58), and Greenstein (1996, 32–33).

6. Simon Choat observes that this quality for Derrida distinguishes deconstruction from structuralism, given the latter's "ambivalence" towards the "ethic of presence" that permeates its conditions (Choat 2010, 14). Choat is here referring to Derrida's engagement of Lévi-Strauss in *Writing and Difference* (Derrida 1978 [1967]), regarding the self-presence by which speech is defined.

7. Derrida also responds to Habermas in another lengthy footnote in *Memoires for Paul de Man* (Derrida 1989, 259–61), in which he focuses on Habermas's accusation that "Derrida does not belong to those philosophers who like to argue" (1989, 260). In this response Derrida outlines his contrary interpretation of the actual nature of philosophical communication that deconstruction offers.

8. In noting Habermas's criticism of the compromised communicability, or "impenetrability," of Derrida's deconstructive method, as well as its blurring of the "law of genre which

divides philosophy and literature," John Elias notes that if "one were to seriously follow Habermas, then we would also have to say that both Nietzsche and Kierkegaard were also non-philosophers who refused to communicate with their readers" (Elias 1998, 61).

9. As Benoît Peeters's *Derrida: A Biography* (2013) describes, there was a reconciliation between Derrida and Habermas in the late 1990s that led to them collaborating over the following few years. Such collaborations include Giovanna Borradori's *Philosophy in a Time of Terror: Dialogues with Jürgen Habermas and Jacques Derrida* (2003), as well as Habermas and Derrida's jointly written chapter on a European response to American foreign policy (Habermas and Derrida 2005).

10. A comparative analysis of Mead's and Derrida's respective models of time is undertaken by Sandra Rosenthal in "Giving Ourselves a Little Time for Mead and for Derrida—and Why Bother" (Rosenthal 1993). It must be noted though that Rosenthal's study differs from mine in being occupied by the topic of an understanding of the self.

11. An interpretation of the relative lack of time available to women after the incursions of first-wave feminism has a heritage in the thought of Julia Kristeva, in particular as found in her essay "Women's Time" (Kristeva 1981).

12. A reference to Aristotle's "On Interpretation," as found in his *The Organon, or Logical Treatises*, in which it is stated that "spoken words are the symbols of mental experience and written words are the symbols of spoken words" (Aristotle 1853, 1.1).

13. See Margo DeMello's *Bodies of Inscription: A Cultural History of the Modern Tattoo Community* (2000, 71–77), which raises Western appropriations of Japanese tattooing aesthetics.

14. This is a fairly common interpretation of the Western perspective of the imagery found in Asian tattooing styles. For one such commentary, see Hardy (1995).

15. *The Ethics of Deconstruction: Derrida and Levinas* (1999) by Simon Critchley attempts to assimilate Derrida's originary difference with Levinas's ethical metaphysics. This inspires counter arguments, such as Martin Hägglund's "The Necessity of Discrimination: Disjoining Derrida and Levinas" (2004), which demands that such assimilation incompatibly reduces deconstruction to a series of non-violent relations.

5. MATERIAL CLIMATES, MATERIAL THEORIES

1. See, in particular, chapter 5: "Signs and the Blink of an Eye" (Derrida 1973, 60–69).

2. Sawicki (1997, 112–13); Paterson (2007, 22–23). As Russell (2006) also notes, the "nature of the link" between Husserl's presentation of the body and the psyche in *Ideas II* is "one of the most hotly debated topics in philosophy" (2006, 153).

3. While Husserl admires Descartes's interrogation of the certainty of physical externality, describing him in *Cartesian Meditations* as "France's greatest thinker" (Husserl 1999, 1), he considers mind-body dualism to be a pseudo-problem. This, in Husserl's opinion, is because Descartes does not take the issue of doubt far enough and "fails to see its true significance, that of transcendental subjectivity" (Husserl 1964 [1929], 9).

4. See Hegel's *Phenomenology of Spirit* (1977 [1807]), where Hegel commentates in the preface that "Kant rediscovered this triadic form by instinct" (Hegel 1977, S50).

5. Smyth (2013, 135); Matthews (2014, 114).

6. Armed with these insights, we can respond to the kind of critique that Bourdieu makes that phenomenology has too particular a focus. On the one hand, Bourdieu recognizes a scientific appeal in Husserl's and Merleau-Ponty's phenomenologies, acknowledging how they could serve a "desire to shun fashionable enthusiasms and could also lead one to seek another antidote to the 'facile' aspects of existentialism" (Bourdieu 2008, 11). However, Bourdieu's concern as detailed in *Political Interventions: Social Science and Political Action* (2008 [2002]), is that aspects of phenomenology's method (as chiefly exhibited in Sartre's existentialism) potentially neglect to reflect on our relationship with our entire environment by only attending to our most familiar, interpersonal environment. Bourdieu warns that anyone undertaking an "ethno-methodology" (such as phenomenology) must "not present this science of the

lived experience of the social world as the science of the social world" (Bourdieu 2008 [2004], 69). In considering "critical phenomenologies after Merleau-Ponty and Bourdieu," for Jérôme Melançon this point means that phenomenology for Bourdieu is "not radical enough," for it does not "reflect on our familiarity with our environment as a whole" (Melançon 2014, 4). While Merleau-Ponty was not really in Bourdieu's sights, Merleau-Ponty's above engagement with Marx conversely indicates, I suggest, a phenomenological receptivity with a whole environment, where "whole" implies a longitudinal temporality of social history.

7. A document produced by the United Nations Framework Convention on Climate Change, demanding that global warming be arrested by cutting industrially induced emissions (McKibbin, Morris, and Wilcoxen 2011).

8. See Adam (1990, 68); Grosz (1999, 1); and even Durkheim (1952, 270).

9. For a collection of object-oriented ontology's explorations on a range of topics that directly and indirectly appraise the possibility of nonhuman sentience and consciousness, see the Richard Grusin edited *The Nonhuman Turn* (2015).

10. The possibility of the sentience of plants is opened by Peter Tompkins and Christopher Bird in *The Secret Life of Plants* (1973). Anthony Trewavas, professor in plant physiology and molecular biology, later builds on this work in considering whether plants have what we would call intelligence (2003).

11. For one of many discussions on this theme, see Robert Nicholls, Frank Hoozemans, and Marcel Marchand's *Increasing Flood Risk and Wetland Losses Due to Global Sea-Level Rise: Regional and Global Analyses* (1999).

6. METHODS OF ACCOMMODATING LATENESS

1. It is this conception of multiplicity to which Gilles Deleuze is drawn, portraying it in *Bergsonism* (1988 [1966]) as the legacy of Bergson's philosophy.

2. In *Critique of Pure Reason* (1999 [1781]), Kant argues that the world can be known as it appears (is represented), but not absolutely. This, for Bergson, illustrates the aforementioned, socially normative, spatializing tendencies that ensure social function/survival, whereby as stated in *The Creative Mind*, "spatiality, and . . . sociability, are . . . the real causes of the relativity of our knowledge" (Bergson 2007, 16).

3. Grosz (2005, 48); Andrew (2012, 45).

4. Bergson later also describes the "*melting into one another*" of time states, "forming an organic whole" (Bergson 1960, 128; author's original emphasis).

5. Gillies's reading develops the inspired commentary on Bergson's perspective on art offered by Arthur Szathmary's *The Aesthetic Theory of Henri Bergson* (Szathmary 1937). Szathmary particularly addresses Bergson's argument discerning the "vital experience" of art (1937, 50).

6. Thomas Quirk also discusses how this "later critical theory was greatly influenced by Bergsonism and appropriated its subtler, more useful logic" (Quirk 1990, 72). Equally relevant is Oliver Tearle's review of the relationship between Hulme, Davie, Bergson, and New Criticism (Tearle 2013, 107–9).

7. The editors of *Radical Philosophy*, when introducing Horkheimer's essay on Bergson, actually cite the lack of "engagement with established criticisms" (Horkheimer 2005, 9) of Bergson's conception of history as partly motivating their republication of Horkheimer's piece.

8. See also on this theme where Merleau-Ponty states that past time is *wholly* grasped in the present (Merleau-Ponty 1962, 80).

9. As Reynolds acknowledges, this congruence between Derrida and Merleau-Ponty is also discussed in Robert Vallier's essay *Blindness and Invisibility: The Ruins of Self-Portraiture (Derrida's Re-Reading of Merleau-Ponty)* (Vallier 1997).

10. Deleuze reiterates Bergson's demand that recollections and times of which we are not presently aware still exist, like "the actual existence of non-perceived objects in space" (Deleuze 1989, 80).

11. Deleuze not only states that "the Bergsonian method has shown two main aspects, the one dualist, the other monist" (Deleuze 1988, 73), but also that "starting from monism, we are able to rediscover dualism and account for it" (1988, 94).

12. Deleuze's interpretation is often incorporated straightforwardly (Bogue [2003]; Somers-Hall [2012]; Kebede [2016]) in order to avoid an exclusively monist or dualist reading of Bergson. Dosse (2011, 138–40) conversely argues that Bergson's position, and Deleuze's reading of Bergson, leans toward a monism.

Bibliography

Adam, Barbara. 1990. *Time and Social Theory*. Cambridge: Polity Press.

Adam, Barbara. 1993. "Time and Environmental Crisis: An Exploration with Special Reference to Pollution." *Innovation: The European Journal of Social Science Research* 6 (4): 399–413.

Adam, Barbara. 1994. "Running Out of Time: Global Crisis in Human Management." In *Social Theory and the Global Environment*, edited by Michael Redcliff and Ted Benton, 99–112. London: Routledge.

Adam, Barbara. 1995. *Timewatch: The Social Analysis of Time*. Cambridge: Polity Press.

Adjaye, Joseph. 1994. "Time in Africa and Its Diaspora: An Introduction." In *Time in the Black Experience*, edited by Joseph Adjaye, 1–16. Westport: Greenwood Publishing Group.

Adorno, Theodor. 2001. *The Culture Industry: Selected Essays on Mass Culture*. Edited by J. M. Bernstein. London and New York: Routledge.

Alcocer, Rudyard. 2012. *Time Travel in the Latin American and Caribbean Imagination*. New York: Palgrave MacMillan.

Amiel, Henri-Frédéric. 2005 (1882). *Amiel's Journal*. Translated by Humphry Ward. Cirencester: Echo Library Publications.

Anaximander. 1983 (circa 550 B.C.E.). "Anaximander of Miletus." In *The Presocratic Philosophers: A Critical History with a Selection of Texts*, edited by Geoffrey Kirk, John Raven, and Malcolm Schofield, 99–142. Cambridge: Cambridge University Press.

Andreason, Robin. 2004. "The Cladistic Race Concept: A Defense." *Biology and Philosophy* 19 (3): 425–42.

Andrew, Patricia. 2012. *The Social Construction of Age: Adult Foreign Language Learners*. Bristol, Buffalo, Toronto: Multilingual Matters.

Anshelm, Jonas, and Anders Hansson. 2014. "The Last Chance to Save the Planet? An Analysis of the Geoengineering Advocacy Discourse in the Public Debate." *Environmental Humanities* 5 (1): 101–23.

Appiah, Kwame Anthony. 1996. "Part 1. Analysis: Against races." In *Colour Conscious: The Political Morality of Race*, edited by Kwame Anthony Appiah and Amy Gutmann, 30–73. Princeton: Princeton University Press.

Aristotle. 1853. *The Organon, or Logical Treatises, of Aristotle: Volume II*. Translated by Octavius Freire Owen, M.A. Covent Garden: Henry Bohn.

Aristotle. 1996. *Physics*. Translated by Robin Waterfield. Oxford and New York. Oxford University Press.

Aristotle and Porphyry. 2010 (circa 50 B.C.E.). *The Organon, or Logical Treatises, of Aristotle Volume 1*. Translated by Octavius Freire Owen, M.A. Charlestown: BiblioBazaar.

Arriaga, Manuel. 2006. *The Modernist-Postmodernist Quarrel on Philosophy and Justice: A Possible Levinasian Meditation*. Lanham: Lexington Books.
Augustine of Hippo. 1961 (circa 400). *Confessions*. Translated by R. Pine-Coffin. London and New York: Penguin Books.
Aveni, Anthony. 1989. *Empires of Time: Calendars, Clocks, and Cultures*. London and New York: Tauris Parke Paperbacks.
Badia, Lynn. 2016. "Theorizing the Social: Émile Durkheim's Theory of Force and Energy." *Cultural Studies* 30 (6): 969–1000.
Baert, Patrick, and Filipe Carreira da Silva. 2010. *Social Theory in the Twentieth Century and Beyond*. 2nd ed. Cambridge and Malden: Polity Press.
Baldick, Chris. 2014. *Criticism and Literary Theory: 1890 to the Present*. London and New York: Routledge.
Barad, Karen. 2007. *Meeting the Universe Halfway: Quantum Physics and the Entanglement of Matter and Meaning*. Durham and New York: Duke University Press.
Bardon, Adrian. 2013. *A Brief History of the Philosophy of Time*. Oxford: Oxford University Press.
Barrow, Isaac. 2009 (1683). *Geometrical Lectures*. Translated by James Child. Charlestown: BiblioBazaar.
Barthes, Roland. 1990. *The Fashion System*. Translated by Matthew Ward and Richard Howard. Berkeley, Los Angeles and London: University of California Press.
Bash, Harry. 2000. "A Sense of Time: Temporality and Historicity in Sociological Inquiry." *Time & Society* 9 (2-3): 187–204.
Bell, Philip. 2018. "The Limits of Semiotics—Epistemology and the Concept of 'Race.'" In *Advancing Multimodal and Critical Discourse Studies: Interdisciplinary Research Inspired by Theo Van Leeuwen's Social Semiotics*. Edited by Sumin Zhao, Emilia Djonov, Anders Björkvall, and Morten Boeriis. New York and Oxon: Routledge.
Benveniste, Émile. 1971 (1966). *Problems in General Linguistics*. Translated by Mary Benveniste. Coral Gables: University of Miami Press.
Berger, Peter. 1963. *Invitation to Sociology: A Humanistic Perspective*. New York: Doubleday Books.
Berger, Peter, and Thomas Luckmann. 1966. *The Social Construction of Reality: A Treatise in the Sociology of Knowledge*. New York: Random House.
Bergson, Henri. 1911 (1907). *Creative Evolution*. Translated by Arthur Mitchell. New York: Henry Holt and Company.
Bergson, Henri. 1960 (1889). *Time and Free Will: An Essay on the Immediate Data of Consciousness*. Translated by Frank Pogson. New York: Harper Torchbooks.
Bergson, Henri. 2004 (1908). *Matter and Memory*. Translated by Nancy Paul and W. Scott Palmer. New York: Dover Publications.
Bergson, Henri. 2007 (1934). *The Creative Mind: An Introduction to Metaphysics*. Translated by Mabelle Andison. New York: Dover Publications.
Berman, Art. 1988. *From the New Criticism to Deconstruction: The Reception of Structuralism and Post-Structuralism*. Urbana and Chicago: University of Illinois Press.
Birth, Kevin. 2012. *Objects of Time: How Things Shape Temporality*. New York: Palgrave MacMillan.
Boghossian, Paul. 2006. *Fear of Knowledge: Against Relativism and Constructivism*. Oxford: Oxford University Press.
Bogost, Ian. 2015. "The Aesthetics of Philosophical Carpentry." In *The Nonhuman Turn*, edited by Richard Grusin, 81–100. Minneapolis: University of Minnesota Press.
Bogue, Ronald. 2003. *Deleuze on Cinema*. New York and Oxon: Routledge.
Bohman, James. 2003. "Critical Theory as Practical Knowledge: Participants, Observers, and Critics." In *The Blackwell Guide to the Philosophy of the Social Sciences*, edited by Stephen Turner and Paul Roth, 91–109. Malden and Oxford: Blackwell Publishing.
Boorstin, Daniel. 1985. *The Discoverers*. New York: Random House Inc.
Borradori, Giovanna (Editor). 2003. *Philosophy in a Time of Terror: Dialogues with Jürgen Habermas and Jacques Derrida*. Chicago: University of Chicago Press.

Bourdieu, Pierre. 1963. "The Attitude of the Algerian Peasant toward Time." In *Mediterranean Countrymen: Essays in the Social Anthropology of the Mediterranean*, edited by Julian Pitt-Rivers, 55–72. Paris and the Hague: Mouton & Co.

Bourdieu, Pierre. 1990 (1980). *The Logic of Practice*. Translated by Richard Nice. Cambridge: Polity Press.

Bourdieu, Pierre. 2008 (2002). *Political Interventions: Social Science and Political Action*. Translated by David Fernbach. London and New York: Verso.

Bourdieu, Pierre. 2008 (2004). *Sketch for a Self-Analysis*. Translated by Richard Nice. Chicago and London: The University of Chicago Press.

Boyd, Richard. 1992. "Constructivism, Realism, and Philosophical Method." In *Inference, Explanation, and Other Frustrations: Essays in the Philosophy of Science*, edited by John Earman, 131–98. Berkeley and Los Angeles: University of California Press.

Brain, Robert. 1979. *The Decorated Body*. New York: Harper and Row.

Brame, Gloria, William Brame, and Jon Jacobs. 1993. *Different Loving: An Exploration of the World of Sexual Dominance and Submission*. New York: Villard Books.

Brody, Harold. 1955. "Organization of the Cerebral Cortex: A Study of Aging in the Human Cerebral Cortex." *The Journal of Comparative Neurology* 102 (2): 511–56.

Bronner, Stephen, and Douglas Kellner. 1989. "Introduction." In *Critical Theory and Society: A Reader*, edited by Stephen Bronner and Douglas Kellner, 1–21. London and New York: Routledge.

Brooks, Ann. 1997. *Postfeminisms: Feminism, Cultural Theory and Cultural Forms*. London and New York: Routledge.

Brown, Donald. 2010. "A Comprehensive Ethical Analysis of the Copenhagen Accord." https:/ /ethicsandclimate.org/2010/01/31/a-comprehensive-ethical-analysis-of-the-copenhagen-accord/.

Brown, Norman. 1990. "Organization of Public Events in Long-Term Memory." *Journal of Experimental Psychology: General* 119 (3): 297–314.

Brown, Ted. 1992. "Progeria: A Human-Disease Model of Accelerated Aging." *American Journal of Clinical Nutrition* 55 (6): 1222S–24S.

Burchard, Esteban, Elad Ziv, Natasha Coyle, Scarlett Lin Gomez, Hua Tang, Andrew Karter, Joanna Mountain, Elisio Pérez-Stable, Dean Sheppard, and Neil Risch. 2003. "The Importance of Race and Ethnic Background in Biomedical Research and Clinical Practice." *The New England Journal of Medicine* 348: 1170–75.

Burt, Stephen, and Jennifer Lewin. 2003. "Poetry and the New Criticism." In *A Companion to Twentieth-Century Poetry*, edited by Neil Roberts, 153–67. Malden: Blackwell Publishing.

Butler, Judith. 1990. *Gender Trouble: Feminism and the Subversion of Identity*. New York and London: Routledge.

Campos, Milton. 2009. "Critical Constructivism." In *Encyclopedia of Communication Theory*, edited by Stephen Littlejohn and Karen Foss, 216–19. London, New Delhi, Singapore, and Thousand Oaks: Sage Publications.

Cannon, Kathleen. 2012. *She Said What? Quotable Women Talk Aging*. Minneapolis-Saint Paul: The Cannon Agency.

Carrington, Damien. 2013. "Global Carbon Dioxide in Atmosphere Passes Milestone Level." *The Guardian*, May 10, 2013. http://www.theguardian.com/environment/2013/may/10/carbon-dioxide-highest-level-greenhouse-gas.

Carroll, Sean. 2010. *From Eternity to Here: The Quest for the Ultimate Theory of Time*. New York: Dutton.

Cheah, Pheng, 2010. "Non-Dialectical Materialism." In *New Materialisms: Ontology, Agency, and Politics*, edited by Diana Coole and Samantha Frost, 70–91. Durham and London: Duke University Press.

Chimisso, Cristina, and Gad Freudenthal. 2003. "A Mind of Her Own: Hélène Metzger to Émile Meyerson, 1933." *Isis* 94 (3): 477–91.

Choat, Simon. 2010. *Marx through Post-Structuralism: Lyotard, Derrida, Foucault, Deleuze*. London and New York: Continuum International Publishing Group.

Cohen, Richard. 1999. "Philo, Spinoza, Bergson: The Rise of an Ecological Age." In *The New Bergson*, edited by John Mullarkey, 18–31. Manchester and New York: Manchester University Press.

Cohen, Stephen. 2002. "Between Form and Culture: New Historicism and the Promise of a Historical Formalism." In *Renaissance Literature and Its Formal Engagements*, edited by Mark Rasmussen, 17–42. New York and Hampshire: Palgrave.

Cohn, Marc. 2007. *The Mathematics of the Calendar*. North Carolina: Lulu.

Collins, Guy. 2015. *Faithful Doubt: The Wisdom of Uncertainty*. Cambridge: The Lutterworth Press.

Collins, Randall. 2007. "The Classical Tradition in Sociology of Religion." In *The Sage Handbook of the Sociology of Religion*, edited by James Beckford and N. J. Demerath III, 19–38. London, Thousand Oaks, New Delhi, and Singapore: Sage Publications Ltd.

Coole, Diana. 2007. *Merleau-Ponty and Modern Politics after Anti-Humanism*. Maryland: Rowman & Littlefield Publishers, Inc.

Coveney, Peter, and Roger Highfield. 1990. *The Arrow of Time*. London: Flamingo.

Critchley, Simon. 1999. *The Ethics of Deconstruction: Derrida and Levinas*. West Lafayette: Purdue University Press.

Crossley, Nick. 1995. "Merleau-Ponty, the Elusive Body and Carnal Sociology." *Body and Society* 1 (1): 43–63.

Crossley, Nick. 2001. "Embodiment and Social Structure: A Response to Howson and Inglis." *The Sociological Review* 49 (3): 318–26.

Crossley, Nick. 2005. "Sociology and the Body." In *Sage Handbook of Sociology*, edited by Craig Calhoun, Chris Rojek, and Bryan Turner, 442–56. London, New Delhi, and Thousand Oaks: Sage Publications.

Davie, Donald. 1976. *Articulate Energy: An Inquiry into the Syntax of English Poetry*. London: Routledge and Keegan.

Daylight, Russell. 2011. *What If Derrida Was Wrong about Saussure?* Edinburgh: Edinburgh University Press.

Deleuze, Gilles. 1988 (1966). *Bergsonism*. Translated by Hugh Tomlinson and Barbara Habberjam. New York: Zone Books.

Deleuze, Gilles. 1989 (1985). *Cinema 2: The Time Image*. Translated by Hugh Tomlinson and Robert Galeta. London: The Athlone Press.

DeMello, Margo. 2000. *Bodies of Inscription: A Cultural History of the Modern Tattoo Community*. Durham: Duke University Press.

DeMello, Margo. 2007. *Encyclopedia of Body Adornment*. London and Westport: Greenwood Press.

Derrida, Jacques. 1973 (1967). *Speech and Phenomena, and Other Essays on Husserl's Theory of Signs*. Translated by David Allison. Evanston: Northwestern University Press.

Derrida, Jacques. 1976 (1967). *Of Grammatology*. Translated by Gayatri Spivak. Baltimore: The John Hopkins University Press.

Derrida, Jacques. 1977. *Limited Inc*. Evanston: Northwestern University Press.

Derrida, Jacques. 1978 (1967). *Writing and Difference*. Translated by Alan Bass. Chicago: University of Chicago Press.

Derrida, Jacques. 1981 (1972). *Positions*. Translated by Alan Bass. Chicago: University of Chicago Press.

Derrida, Jacques. 1982 (1972). *Margins of Philosophy*. Translated by Alan Bass. Chicago: University of Chicago Press.

Derrida, Jacques. 1985. "Letter to a Japanese Friend." In *Derrida and Difference*, edited by David Wood and Robert Bernasconi, 1–6. Translated by David Wood and Andrew Benjamin. Evanston: Northwestern University Press.

Derrida, Jacques. 1989. *Memoires for Paul de Man*. Translated by Cecile Lindsay, Jonathan Culler, Eduardo Cadava, and Peggy Kamuf. New York and Oxford: Columbia University Press.

Derrida, Jacques. 1993. *Memoirs of the Blind: The Self-Portrait and Other Ruins*. Translated by Pascale-Anne Brault and Michael Naas. Chicago: University of Chicago Press.

Dosse, François. 2011. *Gilles Deleuze and Félix Guattari: Intersecting Lives*. Translated by Deborah Glassman. New York: Columbia University Press.

Douglas, Jack. 1967. *The Social Meanings of Suicide*. Princeton: Princeton University Press.

Durkheim, Émile. 1915 (1912). *The Elementary Forms of Religious Life*. Translated by Joseph Swain. London: Allen and Unwin.

Durkheim, Émile. 1938 (1895). *The Rules of Sociological Method*. Edited by George Catlin. Translated by Sarah Solovay and John Mueller. New York: The Free Press.

Durkheim, Émile. 1952 (1897). *Suicide: A Study in Sociology*. Edited by George Simpson. Translated by John Spaulding and George Simpson. London and New York: Routledge.

Durkheim, Émile. 1974 (1898). "Individual and Collective Representations." In *Sociology and Philosophy*, edited and translated by David Pocock, 1–34. New York: The Free Press.

Durkheim, Émile. 1995 (1912). *The Elementary Forms of Religious Life*. Translated by Karen Fields. New York: Simon and Schuster.

Edwards, Paul. 2006. "The Imagery of Hulme's Poems and Notebooks." In *T. E. Hulme and the Questions of Modernism*, edited by Edward Comentale and Andrzej Gasiorek, 23–38. Aldershot and Burlington: Ashgate.

Elder-Vass, Dave. 2012. *The Reality of Social Construction*. Cambridge: Cambridge University Press.

Einstein, Albert. 2006 (1920). *Relativity: The Special and the General Theory*. Edited by Nigel Calder. Translated by Robert Lawson. London and New York: Penguin Books.

Elias, John. 1998. *Philosophical Notes to My Friends*. Toronto, Buffalo, and Lancaster: Guernica.

Elias, Norbert. 1992 (1987). *Time: An Essay*. Oxford and Cambridge: Blackwell Publishers.

Engammare, Max, and Karin Maag. 2009. *On Time, Punctuality, and Discipline in Early Modern Calvinism*. Cambridge: Cambridge University Press.

Engel, Albert. 1969. "Time and the Earth." *American Scientist* 57 (4): 458–83.

Evans-Pritchard, Edward. 1940. *The Nuer*. Oxford: Oxford University Press.

Featherstone, Mike. 2000. "Body Modification: An Introduction." In *Body Modification*, edited by Mike Featherstone, 1–13. London and Thousand Oaks: Sage Publications.

Feenberg, Andrew. 1999. *Questioning Technology*. London and New York: Routledge.

Felski, Rita. 2000. *Doing Time: Feminist Theory and Postmodern Culture*. New York: New York University Press.

Fish, Jonathan. 2017. *Defending the Durkheimian Tradition: Religion, Emotion and Morality*. London and New York: Routledge.

Flaherty, Michael. 1991. "The Perception of Time and Situated Engrossment." *Social Psychology Quarterly* 54 (1): 76–85.

Flaherty, Michael. 1999. *A Watched Pot: How We Experience Time*. New York: New York University Press.

Flaherty, Michael. 2011. *The Textures of Time*. Philadelphia: Temple University Press.

Flewelling, Colleen. 2005. *The Social Relevance of Philosophy: The Debate over the Applicability of Philosophy to Citizenship*. Lanham and Oxford: Lexington Books.

Forman, Frieda. 1985. "Reflections on Feminizing Time." *Canadian Women's Studies* 6 (2): 27–32.

Forman, Frieda, and Caoran Sowton. 1989. *Taking Our Time: Feminist Perspectives on Temporality*. Oxford: Pergamon Press.

Foucault, Michel. 1977 (1975). *Discipline and Punish: The Birth of the Prison*. Translated by Alan Sheridan. London and New York: Penguin Books.

Foucault, Michel. 1978 (1976). *The History of Sexuality, Volume 1: An Introduction*. Translated by Robert Hurley. New York: Pantheon.

Foucault, Michel. 1980. *Power/Knowledge*. Translated by Colin Gordon. New York: Pantheon.

Fowles, Jib. 1974. "On Chronocentrism." *Futures* 6 (1): 65–68.

Freud, Sigmund. 1989 (1930). *Civilization and Its Discontents*. Edited and translated by James Strachey. London and New York: W. W. Norton and Company Inc.

Fu-Kiau, K. K. Bunseki. 1994. "Ntangu-Tandu-Kolo: The Bantu-Kongo Concept of Time." In *Time in the Black Experience*, edited by Joseph Adjaye, 17–34. Westport: Greenwood Publishing Group.

Galilei, Galileo. 1914 (1632). *Dialogues Concerning Two New Sciences*. Translated by Henry Crew and Alfonso De Salvio. New York: Dover Publications Inc.
Gans, Eric. 2000. "The Body Sacrificial." In *The Body Aesthetic: From Fine Art to Body Modification*, edited by Tobin Siebers, 159–78. Ann Arbor: The University of Michigan Press.
Geldard, Richard. 2000. *Remembering Heraclitus*. Herndon: Lindisfarne Books.
Gell, Alfred. 1993. *Wrapping in Images: Tattooing in Polynesia*. Oxford: Oxford University Press.
Gerber, Timofei. 2017. "Bergson - Philosophy as Attention." In *Epoché: Philosophy Monthly* 3 (June). https://epochemagazine.org/bergson-philosophy-as-attention-eacd6e6074fc.
Giddens, Anthony. 1984. *The Constitution of Society: Outline of the Theory of Structuration*. Cambridge: Polity Press.
Gillies, Mary. 1996. *Henri Bergson and British Modernism*. Montreal, Kingston, London, and Buffalo: McGill-Queen's University Press.
Gingrich, Heather. 2006. "An Examination of Dissociative Symptoms as They Relate to Indigenous Filipino Concepts." *Social Science Diliman* 3 (1–2): 1–48.
Greenstein, Edward. 1996. "Deconstruction and the Biblical Narrative." *Interpreting Judaism in a Postmodern Age*, edited by Stephen Kepnes, 21–54. New York and London: New York University Press.
Gross, David. 1985. "Temporality and the Modern State." *Theory and Society* 14 (1): 53–82.
Gross, Paul, and Norman Levitt. 1994. *Higher Superstition: The Academic Left and Its Quarrels with Science*. Baltimore: Johns Hopkins University Press.
Grosz, Elizabeth. 1998. "The Time of Violence: Deconstruction and Value." *Cultural Values* 2 (2-3): 190–205.
Grosz, Elizabeth. 1999. "Becoming . . . an Introduction." *Becomings: Explorations in Time, Memory and Futures*, edited by Elizabeth Grosz, 1–12. Ithica and London: Cornell University Press.
Grosz, Elizabeth. 2005. *Time Travels: Feminism, Nature, Power*. Crows Nest, N. S. W.: Allen & Unwin.
Grusin, Richard, ed. 2015. *The Nonhuman Turn*. Minneapolis: University of Minnesota Press.
Guia, Katrin de. 2000. "Indigenous Values for Sustainable Nation Building." *Prajna Vihara* 14 (1-2): 175–92.
Gustafson, Darren. 2012. "Jeepney Spirituality." *Thesis Eleven* 112 (1): 87–97.
Habermas, Jürgen. 1984 (1981). *The Theory of Communicative Action: Volume 1: Reason and the Rationalization of Society*. Translated by Thomas McCarthy. Boston: Beacon Press.
Habermas, Jürgen. 1987 (1985). *The Philosophical Discourse of Modernity: Twelve Lectures*. Translated by Frederick Lawrence. Cambridge: Polity Press.
Habermas, Jürgen. 1991 (1962). *The Structural Transformation of the Public Sphere: An Inquiry into a Category of Bourgeois Society*. Translated by Thomas Burger and Frederick Lawrence. Cambridge: The MIT Press.
Habermas, Jürgen, and Jacques Derrida. 2005. "February 15, or, What Binds Europeans Together: Plea for a Common Foreign Policy, Beginning in Core Europe." In *Old Europe, New Europe, Core Europe: Transatlantic Relations after the Iraq War*, edited by Daniel Levy, Max Pensky, and John Torpey, 3–13. London and New York: Verso.
Hacking, Ian. 1999. *The Social Construction of What?* Cambridge: Harvard University Press.
Hägglund, Martin. 2004. "The Necessity of Discrimination: Disjoining Derrida and Levinas." *Diacritics* 34 (1): 40–71.
Halbwachs, Maurice. 1992 (1941). *On Collective Memory*. Translated by Lewis Coser. Chicago: The University of Chicago Press.
Halewood, Michael. 2014. *Rethinking the Social through Durkheim, Marx, Weber and Whitehead*. London and New York: Anthem Press.
Hannam, Peter. 2013. "Global Carbon Dioxide Levels Hit Landmark High." *The Sydney Morning Herald*, May 11, 2013. http://www.smh.com.au/environment/climate-change/global-carbon-dioxide-levels-hit-landmark-high-20130510-2je8u.html.
Hansen, Nick. 2009. *Storms of My Grandchildren: The Truth about the Coming Climate Catastrophe and Our Last Chance to Save Humanity*. London: Bloomsbury Publishing.

Haraway, Donna. 1985. "Manifesto for Cyborgs: Science, Technology, and Socialist Feminism in the 1980s." *Socialist Review* 15 (2): 65–108.
Haraway, Donna. 2004. *The Haraway Reader*. London and New York: Routledge.
Hardy, Don Ed. 1995. *Pierced Hearts and True Love: A Century of Drawings for Tattoos*. New York: The Drawing Center.
Harman, Graham. 2002. *Tool-Being: Heidegger and the Metaphysics of Objects*. Chicago and La Salle: Open Court.
Harvey, David. 1990. "Between Space and Time: Reflections on the Geographical Imagination." *Annals of the Association of American Geographers* 80 (3): 418–34. London and New York: Routledge.
Haslanger, Sally. 2012. *Resisting Reality: Social Construction and Social Critique*. New York: Oxford University Press.
Hegel, Georg. 1977 (1807). *Phenomenology of Spirit*. Translated by A. Miller. Oxford: Oxford University Press.
Heidegger, Martin. 1962 (1927). *Being and Time*. Translated by John Macquarrie and Edward Robinson. New York: Harper Collins.
Heidegger, Martin. 1962 (1929). *Kant and the Problem of Metaphysics*. Translated by James Churchill. Bloomington: Indiana University Press.
Heidegger, Martin. 2001. *Zollikon Seminars*. Translated by Franz Mayr and Richard Askay. Evanston: Northwestern University Press.
Heidegger, Martin. 2010 (1927). *Being and Time*. Translated by Joan Stambaugh. Albany: State University of New York Press.
Herbert Kogler, Hans. 2011. "Overcoming Semiotic Structuralism: Language and Habitus in Bourdieu." In *The Legacy of Pierre Bourdieu: Critical Essays*, edited by Simon Susen and Bryan Turner, 271–300. London, New York, and Delhi: Anthem Press.
Horkheimer, Max. 1972 (1968). *Critical Theory: Selected Essays*. Translated by Matthew O'Connell. New York: Continuum.
Horkheimer, Max. 2005. "On Bergson's Metaphysics of Time." *Radical Philosophy: Philosophical Journal of the Independent Left* 131 (May/June): 9–19.
Horkheimer, Max, and Theodor Adorno. 2002 (1947). *Dialectic of Enlightenment: Philosophical Fragments*. Edited by Gunzelin Schmid Noerr. Translated by Edmund Jephcott. Stanford: Stanford University Press.
Hoy, David. 2009. *The Time of Our Lives: A Critical History of Temporality*. Cambridge and London: The MIT Press.
Hulme, Thomas. 1936. *Speculations: Essays on Humanism and the Philosophy of Art*. Edited by Herbert Read. London: Kegan Paul, Trench, Trubner & Co. Ltd.
Hulme, Thomas. 1994. *The Collected Writings of T. E. Hulme*. Edited by Karen Csengeri. Oxford: Clarendon Press.
Husserl, Edmund. 1964 (1929). *The Paris Lectures*. Translated by Peter Koestenbaum. The Hague and Boston: Martinus Nijhoff.
Husserl, Edmund. 1983 (1913). *Ideas Pertaining to a Pure Phenomenology and to a Phenomenological Philosophy: First Book: General Introduction to a Pure Phenomenology*. Translated by Fred Kersten. The Hague and Boston: Martinus Nijhoff.
Husserl, Edmund. 1990 (1952). *Ideas Pertaining to a Pure Phenomenology and to a Phenomenological Philosophy: Second Book: Studies in the Phenomenology of Constitution*. Edited and translated by Richard Rojcewicz and André Schuwer. New York: Springer Publishing.
Husserl, Edmund. 1999 (1931). *Cartesian Meditations: An Introduction to Phenomenology*. Translated by Dorion Cairns. Rotterdam: Kluwer Academic Publishers.
Husserl, Edmund. 2001 (1929). *Logical Investigations: Volume 2*. Edited by Dermot Moran. Translated by John Findlay. London and New York: Routledge.
Irigaray, Luce. 1993 (1984). *An Ethics of Sexual Difference*. Translated by Carolyn Burke and Gillian Gill. New York: Cornell University Press.
Jancovich, Mark. 1993. *The Cultural Politics of the New Criticism*. Cambridge, New York, and Melbourne: Cambridge University Press.

Joaquin, Nick. 1982. "Culture as History: The Filipino Soul." In *Asian and Western Writers in Dialogue: New Cultural Identities*, edited by Guy Amirthanayagam, 159–90. London and Basingstoke: The Macmillan Press.
Johncock, Will. 2012. "Modifying the Modifier: Body Modification as Social Incarnation." *Journal for the Theory of Social Behaviour* 42 (3): 241–59.
Johncock, Will. 2017. "Climate Change, Socially Synchronised: Are We Really Running Out of Time?" In *What If Culture Was Nature All Along*, edited by Vicki Kirby, 199–222. Edinburgh: Edinburgh University Press.
Johnson, Walter. 2000. "Possible Pasts: Some Speculations on Time, Temporality, and the History of the Atlantic Slave Trade." *Amerkiastudien/American Studies* 45 (4): 485–99.
Kant, Immanuel. 1999 (1781). *Critique of Pure Reason*. Edited and translated by Paul Guyer and Allen Wood. Cambridge: Cambridge University Press.
Kebede, Messay. 2016. "Beyond Dualism and Monism: Bergson's Slanted Being." *Journal of French and Francophone Philosophy* 24 (2): 106–30.
Kelly, George. 1991. *The Psychology of Personal Constructs: Volume 1: Theory and Personality*. London and New York: Routledge.
Kirby, Vicki. 1997. *Telling Flesh: The Substance of the Corporeal*. London and New York: Routledge.
Kirby, Vicki. 2006. *Judith Butler: Live Theory*. London and New York: Continuum.
Kirby, Vicki. 2011. *Quantum Anthropologies: Life at Large*. Durham: Duke University Press.
Kirksey, Eben. 2015. *Emergent Ecologies*. Durham: Duke University Press.
Klesse, Christian. 1999. "Modern Primivitism: Non-Mainstream Body Modification and Racialized Representation." *Body and Society* 5 (2-3): 15–38.
Kristeva, Julia. 1981. "Women's Time." Translated by Alice Jardine and Harry Blake. *Signs* 7 (1): 13–35.
Laclau, Ernesto. 1990. *New Reflections on the Revolution of Our Time*. London: Verso.
Lai, Chung-Hsiung. 2003. "On Violence, Justice and Deconstruction." *Concentric: Studies in English Literature and Linguistics* 29 (1): 23–46.
Landes, David. 1983. *Revolution in Time*. Cambridge: Harvard University Press.
Larratt, Shannon. 2004. "Shooting the Messenger: Why It's Important to Let Young People Cut." *BME—Publisher's Ring*, 26 July. http://news.bme.com/2004/06/26/why-its-important-to-let-young-people-cut-the-publishers-ring/.
Latour, Bruno. 1993. *We Have Never Been Modern*. Cambridge: Harvard University Press.
Latour, Bruno. 2003. "The Promises of Constructivism." In *Chasing Technoscience: Matrix for Materiality*. Edited by Don Idhe and Evan Selinger. Bloomington: Indiana University Press.
Latour, Bruno, and Steve Woolgar. 1979. *Laboratory Life: The Social Construction of Scientific Facts*. Beverly Hills: Sage Publications.
Lawlor, Leonard. 2003. *Thinking through French Philosophy: The Being of the Question*. Bloomington: Indiana University Press.
Lentricchia, Frank. 1980. *After the New Criticism*. Chicago: The University of Chicago Press.
Levinas, Emmanuel. 1969 (1961). *Totality and Infinity*. Translated by Alphonso Lingis. Pittsburgh: Duquesne University Press.
Levinas, Emmanuel. 1998 (1978). *Otherwise than Being, or Beyond Essence*. Translated by Alphonso Lingis. Pittsburgh: Duquesne University Press.
Lévi-Strauss, Claude. 1992 (1955). *Tristes Tropiques*. Translated by John Weightman and Doreen Weightman. London and New York: Penguin Books.
Lévy-Bruhl, Lucien. 1923 (1922). *Primitive Mentality*. Translated by Lilian Clare. London: George Allen & Unwin Ltd.
Lloyd, Moya. 2013. *Judith Butler: From Norms to Politics*. Cambridge and Malden: Polity Press.
Long, Anthony, and David Sedley. (Editors and translators). 1987. *The Hellenistic Philosophers: Volume 1: Translations of the Principal Sources, with Philosophical Commentary*. Cambridge: Cambridge University Press.
Loseke, Donileen. 2017. *Thinking about Social Problems: An Introduction to Constructionist Perspectives*. 2nd. ed. London and New York: Routledge.
Lucy, Niall. 2004. *A Derrida Dictionary*. Oxford: Blackwell Publishing.

Maggay, Melba. 1998. "Towards Sensitive Engagement with Filipino Indigenous Consciousness." *International Review of Mission* 87 (346): 361–73.

Mahar, Cheleen, Richard Harker, and Chris Wilkes. 1990. "The Basic Theoretical Position." In *An Introduction to the Work of Pierre Bourdieu*, edited by Richard Harker, Cheleen Mahar, and Chris Wilkes, 1–25. London and New York: The MacMillan Press.

Marcuse, Herbert. 1972. *Counter-Revolution and Revolt*. Boston: Beacon Press.

Marcuse, Herbert. 2002 (1964). *One-Dimensional Man: Studies in the Ideology of Advanced Industrial Society*. London and New York: Routledge.

Marx, Karl. 1973. *Economic and Philosophic Manuscripts of 1844 of Karl Marx*. Translated by Martin Milligan. New York: Prometheus Books.

Marx, Karl. 1976 (1867). *Capital: A Critique of Political Economy: Volume One*. Translated by Ben Folkes. Middlesex and New York: Penguin Books.

Massey, Doreen. 1992. "Politics and Space/Time." *New Left Review* 96: 65–84.

Massey, Doreen. 1994. *Space, Place and Gender*. Cambridge: Polity Press.

Matthews, Eric. 2014. *The Philosophy of Merleau-Ponty*. London and New York: Routledge.

Maze, J. 2001. "Social Constructionism, Deconstructionism, and Some Requirements of Discourse." *Theory and Psychology* 11 (3): 393–417.

McCallus, Joseph. 1994. "Discourse Characteristics of a Filipino Electronic Discussion Group." *Philippine Quarterly of Culture and Society* 22 (1): 46–63.

McKibben, William. 2006. *The End of Nature*. New York: Random House.

McKibbin, Warwick, Adele Morris, and Peter Wilcoxen. 2011. "Comparing Climate Commitments: A Model-Based Analysis of the Copenhagen Accord." *Climate Change Economics* 2 (2): 79–103.

Mead, George. 2002 (1932). *The Philosophy of the Present*. New York: Prometheus Books.

Melançon, Jérôme. 2014. "Thinking Corporeally, Socially, and Politically: Critical Phenomenology after Merleau-Ponty and Bourdieu." *Bulletin d'analyse phénoménologique* 10 (8): 1–28.

Merleau-Ponty, Maurice. 1962 (1945). *Phenomenology of Perception*. Translated by Colin Smith. New York and London: Routledge.

Merleau-Ponty, Maurice. 1964 (1961). *Sense and Non-Sense*. Translated by Hubert Dreyfus and Patricia Dreyfus. Evanston: Northwestern University Press.

Merleau-Ponty, Maurice. 1968 (1964). *The Visible and the Invisible: Followed by Working Notes*. Translated by Alphonso Lingis. Evanston: Northwestern University Press.

Merleau-Ponty, Maurice. 1973 (1955). *Adventures of the Dialectic*. Translated by Joseph Bien. Evanston: Northwestern University Press.

Merleau-Ponty, Maurice. 2001 (1968). *The Incarnate Subject: Malebranche, Biran, and Bergson on the Union of Body and Soul*. Edited by Andrew Bjelland Jr. and Patrick Burke. Translated by Paul Milan. New York: Prometheus Books.

Metzger, Hélène. 1987 (1937). "La méthode philosophique dans l'histoire des sciences." In *La méthode philosophique en histoire des sciences: Textes 1914–1939*. Edited by Gad Freudenthal. Paris: Fayard.

Meyer, Robinson. 2015. "Earth's Atmosphere Just Crossed an Epochal Threshold." *The Atlantic*, November 24, 2015. http://www.theatlantic.com/science/archive/2015/11/november-11-2015-the-last-day-of/417378/.

Meyerson, Émile. 1931. *Du cheminement de la pensée*. Paris: Vrin.

Miller, Daniel. 1998. *A Theory of Shopping*. Cambridge: Polity Press.

Misztal, Barbara. 2003. *Theories of Social Remembering*. Maidenhead and Philadelphia: Open University Press.

Moi, Toril. 2017. *Revolution of the Ordinary: Literary Studies after Wittgenstein, Austin, and Cavell*. Chicago and London: University of Chicago Press.

Moore, Wilbert. 1963. *Man, Time and Society*. New York: John Wiley.

Moran, Dermot. 2017. "Intercorporality and Intersubjectivity: A Phenomenological Exploration of Embodiment." In *Embodiment, Enaction, and Culture: Investigating the Constitution of the Shared World*, edited by Christoph Durt, Thomas Fuchs, and Christian Tewes, 25–46. Cambridge and London: MIT Press.

Morton, Timothy. 2013. *Hyperobjects: Philosophy and Ecology after the End of the World*. Minneapolis: University of Minnesota Press.

Nagel, Thomas. 1997. *The Last Word*. Oxford: Oxford University Press.

Nellhaus, Tobin. 1998. "Signs, Social Ontology, and Critical Realism." *Journal for the Theory of Social Behaviour* 28: 1–24.

Newton, Isaac. 1952 (1673). *The Principia: The Mathematical Principles of Natural Philosophy*. Translated by Andrew Motte. Chicago: Encyclopaedia Britannica Inc.

Newton, Isaac, A. Rupert Hall, and Marie Hall. 1978. *Unpublished Scientific Papers of Isaac Newton: A Selection from the Portsmouth Collection in the University Library*. Cambridge: CUP Archive.

Nicholls, Robert, Frank Hoozemans, and Marcel Marchand. 1999. "Increasing Flood Risk and Wetland Losses Due to Global Sea-Level Rise: Regional and Global Analyses." *Global Environmental Change* 9 (1): 69–87.

Nnajiofor, Osita. 2016. "Justification of the Concept of Time in Africa." *OGIRISI: A New Journal of African Studies* 12: 253–81.

Norris, Christopher. 1997. *Against Relativism: Philosophy of Science, Deconstruction, and Critical Theory*. Oxford: Blackwell.

Nowotny, Helga. 1975. "Time Structuring and Time Measurement. On the Interrelation between Timekeepers and Social Time." In *The Study of Time II*, edited by J. T. Fraser and N. Lawrence, 325–42. Berlin: Springer-Verlag.

Nowotny, Helga. 1992. "Time and Social Theory: Towards a Social Theory of Time." *Time & Society* 1 (3): 421–54.

Olkowski, Dorothea. 2010. "In Search of Lost Time: Merleau-Ponty, Bergson, and the Time of Objects." *Continental Philosophy Review* 43 (4): 525–44.

Ortner, Sherry. 1972. "Is Female to Male as Nature Is to Culture?" *Feminist Studies* 1 (2): 5–31.

Oyama, Susan. 2000. *Evolution's Eye: A Systems View of the Biology-Culture Divide*. Durham and London: Duke University Press.

Papert, Seymour. 1987. *A New Opportunity for Science Education*. NSF Grant Application.

Papert, Seymour. 1991. "Situating Constructionism." In *Constructionism: Research Reports and Essays 1985–1990, the Media Lab, Epistemology and Learning Group, Massachusetts Institute of Technology*, edited by Seymour Papert and Idit Harel, 1–18. Norwood: Ablex Publishing Corporation.

Paterson, Mark. 2007. *The Senses of Touch: Haptics, Affects and Technologies*. Oxford and New York: Berg.

Peeters, Benoît. 2013. *Derrida: A Biography*. Cambridge and Malden: Polity Press.

Penley, Constance, and Andrew Ross. 1991. "Cyborgs and Large: Interview with Donna Haraway." In *Technoculture*, edited by Constance Penley and Andrew Ross, 1–26. Minneapolis: University of Minnesota Press.

Peters, Michael, and Gert Biesta. 2009. *Derrida, Deconstruction, and the Politics of Pedagogy*. New York: Peter Lang.

Piaget, Jean. 1952. *The Origins of Intelligence in Children*. New York: International Universities Press Inc.

Piaget, Jean. 1970. *Genetic Epistemology*. Translated by Eleanor Duckworth. New York: W. W. Norton & Company Inc.

Pickering, Andrew. 1984. *Constructing Quarks: A Sociological History of Particle Physics*. Edinburgh: Edinburgh University Press.

Pierrehumbert, Raymond. 2006. "A Catastrophe in Slow Motion." *Chicago Journal of International Law* 6 (2): 573–96.

Pigliucci, Massimo, and Jonathan Kaplan. 2003. "On the Concept of Biological Race and Its Applicability to Humans." *Philosophy of Science* 70 (5): 1161–72.

Pinker, Steven. 2002. *The Blank Slate: The Modern Denial of Human Nature*. London: Penguin Books.

Pitts, Victoria. 2003. *In the Flesh: The Cultural Politics of Body Modification*. Basingstoke and New York: Palgrave Macmillan.

Plato. 2008. *Timaeus*. Translated by Benjamin Jowett. Central: Forgotten Books.

Purnell, Larry, and Betty Paulanka. 2003. "The Purnell Model for Cultural Competence." In *Transcultural Health Care: A Culturally Competent Approach*. 2nd ed. edited by Larry Purnell and Betty Paulanka, 8–39. Philadelphia: F. A. Davis Company.

Quade, Vicki. 1990. "20 Years of Saving the Earth: Barrister Interview with Denis Hayes." *Barrister Magazine*, Fall 1990.

Quirk, Tomas. 1990. *Bergson and American Culture: The Worlds of Willa Cather and Wallace Stevens*. Chapel Hill and London: The University of North Carolina Press.

Rajan, Tilottama. 2002. *Deconstruction and the Remainders of Phenomenology: Sartre, Derrida, Foucault, Baudrillard*. Stanford: Stanford University Press.

Randall, Bryony. 2007. *Modernism, Daily Time and Everyday Life*. Cambridge and New York: Cambridge University Press.

Ransom, John. 1937. "Criticism, Inc." *The Virginia Quarterly Review* 13 (4): 586–602.

Raschke, Carl. 1979. *The Alchemy of the Word: Language and the End of Theology*. Missoula: Scholars Press.

Redford, Donald, ed. 2001. "Cosmetics and Other Body Modifications." In *The Oxford Encyclopaedia of Ancient Egypt: P–Z: Volume 3*, 414–15. Oxford: Oxford University Press.

Reichardt, Ulfried. 2000. "Time and the African-American Experience: The Problem of Chronocentrism." *Amerikastudien/American Studies* 45 (4): 465–84.

Reynolds, Jack. 2004. *Merleau-Ponty and Derrida: Intertwining Embodiment and Alterity*. Athens: Ohio University Press.

Richie, Donald, and Ian Buruma. 1980. *The Japanese Tattoo*. New York: Weatherhill.

Riesman, Daniel. 1950. *The Lonely Crowd: A Study of the Changing American Character*. New Haven and London: Yale University.

Risch, Neil, Esteban Burchard, Elad Ziv, and Hua Tang. 2002. "Categorization of Humans in Biomedical Research: Genes, Race and Disease." *Genome Biology* 3 (7): 1–11.

Rose, Deborah Bird. 2016. "Environmental Humanities." In *Deborah Bird Rose: Love at the Edge of Extinction*. http://deborahbirdrose.com/environmental-humanities/.

Rosenthal, Sandra. 1993. "Giving Ourselves a Little Time for Mead and for Derrida—and Why Bother." *The Journal of Speculative Philosophy* 7 (4): 249–65.

Ross, Andrew. 1996. *The Science Wars*. Durham: Duke University Press.

Runnalls, David. 2008. "Our Common Inaction: Meeting the Call for Institutional Change." *Environment* November/December: 18–29.

Russell, Bertrand. 1961. *A History of Western Philosophy and Its Connection with Political and Social Circumstances from the Earliest Times to the Present Day*. London: George Allen and Unwin Ltd.

Russell, Matheson. 2006. *Husserl: A Guide for the Perplexed*. London and New York: Continuum.

Sanders, Clinton, and D. Angus Vail. 2008. *Customizing the Body: The Art and Culture of Tattooing*. Philadelphia: Temple University Press.

Sartre, Jean-Paul. 2003 (1943). *Being and Nothingness: An Essay on Phenomenological Ontology*. Translated by Hazel Barnes. London and New York: Routledge.

Saussure, Ferdinand de. 1966 (1916). *Course in General Linguistics*. Translated by Wade Baskin. Columbus: McGraw Hill.

Saussure, Ferdinand de. 1972. "From Course in General Linguistics." In *The Structuralists: From Marx to Lévi-Strauss*, edited by Richard de George and Fernande de George, 59–75. New York: Anchor Books.

Sawicki, Marianne. 1997. *Body, Text, and Science: The Literacy of Investigative Practices and the Phenomenology of Edith Stein*. Dordrecht: Springer Science+Business Media.

Scheinfeld, Noah. 2007. "Tattoos and Religion." *Clinics in Dermatology* 25: 362–66.

Schulz, Dirk. 2011. *Setting the Record Queer: Rethinking Oscar Wilde's "The Picture of Dorian Gray" and Virginia Wolf's "Mrs. Dalloway."* Bielefeld: Transcript.

Schütz, Alfred. 1967 (1932). *The Phenomenology of the Social World*. Translated by George Walsh and Frederick Lehnert. Evanston: Northwestern University Press.

Searle, John. 1995. *The Construction of Social Reality*. New York: Free Press.

Searle, Leroy. 2005. "New Criticism." In *The Johns Hopkins Guide to Literary Theory and Criticism*, edited by Michael Groden, Martin Kreiswirth, and Imre Szeman, 691–98. Baltimore: The Johns Hopkins University Press.

Selmer, Jan, and Corinna de Leon. 2002. *Management and Culture in the Philippines*. Hong Kong: Hong Kong Baptist University.

Sesardic, Neven. 2010. "Race: A Social Destruction of a Biological Concept." *Biology and Philosophy* 25 (2): 143–62.

Sheldon, Rebecca. 2015. "Form/Matter/Chora: Object-Oriented Ontology and Feminist New Materialism." In *The Nonhuman Turn*, edited by Richard Grusin, 193–222. Minneapolis: University of Minnesota Press.

Shilling, Chris. 1993. *The Body and Social Theory*. London, New Delhi, and Thousand Oaks: Sage Publications.

Shilling, Chris. 2005. "Embodiment, Emotions and the Foundations of Social Order: Durkheim's Enduring Contribution." In *The Cambridge Companion to Durkheim*, edited by Jeffrey Alexander and Philip Smith, 211–38. Cambridge and New York: Cambridge University Press.

Simmel, Georg. 1971 (1903). "The Metropolis and Mental Life." In *On Individuality and Social Forms*, edited by Donald Levine, 324–40. Chicago and London: University of Chicago Press.

Simmel, Georg. 1997 (1903). "The Metropolis and Modern Life." In *Simmel on Culture*, edited by David Frisby and Mike Featherstone, 174–86. London, Thousand Oaks, and New Delhi: Sage Publications.

Smith, Christian. 2010. *What Is a Person?* Chicago and London: University of Chicago Press.

Smith, Mark. 1997. *Mastered by the Clock: Time, Slavery, and Freedom in the American South*. Chapel Hill: The University of North Carolina Press.

Smyth, Bryan. 2013. *Merleau-Ponty's Existential Phenomenology and the Realization of Philosophy*. London and New York: Bloomsbury.

Snedegar, Keith. 2007. "Problems and Prospects in the Cultural History of South African Astronomy." *African Sky* 11: 27–32.

Snow, Charles. 1998 (1959). *The Two Cultures*. Cambridge: Cambridge University Press.

Sokal, Alan, 1996a. "Transgressing the Boundaries: Toward a Transformative Hermeneutics of Quantum Gravity." *Social Text* 46/47: 217–52.

Sokal, Alan. 1996b. "A Physicist Experiments with Cultural Studies." *Lingua Franca* May-June: 61–64.

Somers-Hall, Henry. 2012. *Hegel, Deleuze, and the Critique of Representation: Dialectics of Negation and Difference*. Albany: State University of New York Press.

Sorokin, Pitrim, and Robert Merton. 1937. "Social Time: A Methodological and Functional Analysis." *The American Journal of Sociology* 42 (5): 615–29.

Stark, Andrew. 2016. *The Consolations of Mortality: Making Sense of Death*. New Haven: Yale University Press.

Stokes, Jane. 2013. *How to Do Media and Cultural Studies*. 2nd ed. Los Angeles, London, Thousand Oaks, New Delhi and Singapore: Sage.

Sullivan, Nikki. 1995. "Illustrative Bodies: Subjectivities, Sociality, Skin Art." *Social Semiotics* 5 (1): 143–58.

Sullivan, Nikki. 2001. *Tattooed Bodies: Subjectivity, Textuality, Ethics and Pleasure*. London and Westport: Praeger.

Sullivan, Nikki. 2004. "Being-Exposed: The Poetics of Sex and Other Matters of Tact." *Transformations* 8 (July). http://www.transformationsjournal.org/journal/issue_08/article_04.shtml.

Sullivan, Nikki. 2009. "The Somatechnics of Bodily Inscription: Tattooing." *Studies in Gender and Sexuality* 10 (3): 129–41.

Sundstrom, Ronald. 2002. "Race as a Human Kind." *Philosophy and Social Criticism* 28 (1): 91–115.

Szathmary, Arthur. 1937. *The Aesthetic Theory of Henri Bergson*. Cambridge: Harvard University Press.

Tabboni, Simonetta. 2001. "The Idea of Social Time in Norbert Elias." *Time & Society* 10 (1): 5–27.

Tallis, Raymond. 1997. *Enemies of Hope: A Critique of Contemporary Pessimism. Irrationalism, Anti-Humanism and Counter-Enlightenment*. Hampshire and New York: Palgrave.

Tearle, Oliver. 2013. *T. E. Hulme and Modernism*. London, New Delhi, New York, and Sydney: Bloomsbury.

Thompson, Ewa. 1971. *Russian Formalism and Anglo-American New Criticism: A Comparative Study*. The Hague and Paris: Mouton.

Tiemersma, Douwe. 1998. "A Model of Organic Time and Development in Africa." *In Temps et developpement dans la pensee de l'Afrique subsaharienne/Time and development in the thought of subsaharan Africa*, edited by Souleymane Bachir Diagne et Heinz Kimmerle, 267–86. Amsterdam: Rodopi.

Toadvine, Ted. 2007. "'Strange Kinship': Merleau-Ponty on the Human-Animal Relation." In *Phenomenology of Life—From the Animal Soul to the Human Mind: Book 1, In Search of Experience*, edited by Anna-Teresa Tymieniecka, 17–32. Dordrecht: Springer.

Tompkins, Peter, and Bird, Christopher. 1973. *The Secret Life of Plants: A Fascinating Account of the Physical, Emotional, and Spiritual Relations between Plants and Man*. New York: Harper and Row Publishers.

Trewavas, Anthony. 2003. "Aspects of Plant Intelligence." *Annals of Botany* 92 (1): 1–20.

Urry, John. 2000. "Sociology of Time and Space." In *The Blackwell Companion to Social Theory*. 2nd ed. edited by Bryan Turner, 416–44. Malden and Oxford: Blackwell Publishers Ltd.

Vallier, Robert. 1997. "Blindness and Invisibility: The Ruins of Self-Portraiture (Derrida's Re-Reading of Merleau-Ponty)." In *Écart and Différance: Merleau-Ponty and Derrida on Seeing and Writing*, edited by Martin Dillon, 191–207. New Jersey: Humanities Press.

Van Der Tuin. 2011. "The New Materialist 'Always Already': On an A-Human Humanities." *Nora-Nordic Journal of Feminist and Gender Research* 19 (4): 285–90.

Van Der Tuin. 2015. "Making Time (for) Duration: Thinking at the Contemporary University." *History of the Present* 5 (2): 187–99.

Vygotsky, Lev. 1986 (1962). *Thought and Language*. Translated by Eugenia Hanfmann and Gertrude Vakar. Cambridge and London: MIT Press.

Wacquant, Loïc. 1992. "Toward a Social Praxeology: The Structure and Logic of Pierre Bourdieu's Sociology." In *An Invitation to Reflexive Sociology*, edited by Pierre Bourdieu and Loïc Wacquant, 1–60. Chicago: University of Chicago Press.

Weber, Max. 2004. *The Vocation Lectures*. Edited by David Owen and Tracy Strong. Translated by Rodney Livingstone. Indianapolis: Hackett Publishing Company.

Weiss, Gail. 1999. *Body Images: Embodiment as Intercorporeality*. London and New York: Routledge.

Weiss, Gail. 2009. "Ambiguity." In *Merleau-Ponty: Key Concepts*, edited by Rosalyn Diprose and Jack Reynolds, 132–41. Durham: Acumen Publishing.

White, David. 2017. *Derrida on Being as Presence*. Berlin and Boston: De Gruyter.

Whitrow, Gerald. 1989. *Time in History*. Oxford: Oxford University Press.

Wright, Terence. 2016. *The Photography Handbook*. 3rd ed. London and New York: Routledge.

Zack, Naomi. 2002. *Philosophy of Science and Race*. New York and Oxon: Routledge.

Zerubavel, Eviatar. 1989. *The Seven Day Circle: The History and Meaning of the Week*. Chicago: University of Chicago Press.

Zerubavel, Eviatar. 2003. *Time Maps: Collective Memory and the Social Shape of the Past*. Chicago: University of Chicago Press.

Index

About the Author

Will Johncock researches structuralist theory, social theory, and Stoic philosophy, with a particular interest in themes regarding both individual and collective times. Recent teaching investigates sociological studies of the patterning of subjectivity and responsibility. Past publications explore embodied relationships with time and distinctions between qualities and quantities of temporal experience.

www.ingramcontent.com/pod-product-compliance
Ingram Content Group UK Ltd.
Pitfield, Milton Keynes, MK11 3LW, UK
UKHW012037150926
475560UK00001B/60

9 781786 611932